Control
of
Polymerization
Reactors

Control
of
Polymerization
Reactors

F. Joseph Schork

School of Chemical Engineering
Georgia Institute of Technology
Atlanta, Georgia

Pradeep B. Deshpande

Department of Chemical Engineering
University of Louisville
Louisville, Kentucky

Kenneth W. Leffew

Engineering Department
E. I. du Pont de Nemours & Company, Inc.
Wilmington, Delaware

With a contribution by
Vikas M. Nadkarni
Chemical Engineering Division
National Chemical Laboratory
Pune, India

CRC Press
Taylor & Francis Group
Boca Raton London New York

CRC Press is an imprint of the
Taylor & Francis Group, an **informa** business

CRC Press
Taylor & Francis Group
6000 Broken Sound Parkway NW, Suite 300
Boca Raton, FL 33487-2742

First issued in paperback 2019

© 1993 by Taylor & Francis Group, LLC
CRC Press is an imprint of Taylor & Francis Group, an Informa business

ISBN-13: 978-0-8247-9043-1 (hbk)
ISBN-13: 978-0-367-40243-3 (pbk)

Library of Congress Cataloging-in-Publication Data

Schork, F. Joseph.
    Control of polymerization reactors / F. Joseph Schork, Pradeep B.
Deshpande, Kenneth W. Leffew, with contributions by Vikas M. Nadkarni.
        p.   cm.
    Includes bibliographical references and index.
    ISBN 0-8247-9043-X (alk. paper)
    1. Polymerization. 2. Chemical process control. I. Deshpande,
Pradeep B. II. Leffew, Kenneth W. III. Title.
TP156.P6S37   1993
660'.28448--dc20   92-44695
                                                                  CIP

Visit the Taylor & Francis Web site at
http://www.taylorandfrancis.com

and the CRC Press Web site at
http://www.crcpress.com

To our wives, Linda (FJS), Meena (PBD), and Debby (KWL),
to our students and teachers, and to Dr. Daniel E. Burke
of Exxon Chemical Company, Baytown, Texas

Preface

The field of synthetic polymers has grown tremendously over the last several decades. Today, polymers are found in a large variety of products—e.g., automobiles, paints, and clothing, to name a few. Polymers have replaced metals in many instances, and with the development of polymer alloys, applications in specialty areas are certain to grow. The new and highly specialized application of polymers, along with the trend toward total quality management and global competitiveness, has served to drive up the quality expectations of the customer. These developments make it imperative to operate the polymerization processes efficiently, which underscores the importance of optimizing controls.

The motivation for this text is the current paucity of books that deal with automatic control of polymerization reactors. An important feature of this work is that it offers the combined treatment of the principles of polymerization reaction engineering *and* the automatic control concepts for polymerization reactors. It is intended to serve as a text for an advanced undergraduate or graduate level course in the chemical engineering curriculum. Instructors should find the text particularly useful because it offers a combined treatment of the steady-state and control aspects of polymerization processes; it is often difficult to accommodate two or more different courses in any one specialty, such as polymerization, into the graduate curriculum. The book should also serve as a valuable reference for engineers in industry.

The first four chapters are concerned with the fundamentals of polymerization kinetics and reactors; the rest of the book is devoted to the study of control strategies for batch and continuous polymerization reactors.

We begin in Chapter 1 with a very brief review of the principles of polymer science. Here we introduce the concept of macromolecules and molecular weights of polymers, followed by a discussion of property–structure correlations of thermoplastics and thermosetting polymers and a description of the methods for analyzing the various properties of polymers. We follow this with a discussion of the chemistry and kinetics of various polymerization reactions. The list includes step-growth polymerization, free radical polymerization, anionic polymerization, and cationic polymerization.

Chapter 2 is devoted to mathematical modeling of polymerization reactors, and the kinetic analysis of anionic, free radical, and step-growth homopolymerization and copolymerization reactions. Chapter 3 deals with polymerization reaction engineering. In this chapter, we study the selection of the type of reactor for the manufacture of a selected polymer; batch, semibatch, continuous stirred tank, and plug flow reactors are considered. We presume a background appropriate to the engineer who has had a single undergraduate course in reactor design. Chapter 4 is devoted to heterogenous polymerization. Here, the processes and kinetics of suspension polymerization, emulsion polymerization, and coordination polymerization are covered.

The first four chapters contain many examples that illustrate the concepts. Chapters 2, 3, and 4 form an introduction to polymerization reaction engineering. Whereas their brevity makes them a less than perfect text from which to teach a complete course in polymerization reaction engineering, they are included for a number of reasons. First, we expect the background of the readers to be varied. Even those who have had a course in polymer science may not have had much exposure to polymerization reaction engineering. Second, we wish to use these chapters to lay the groundwork in the unique features of polymerization reactors that result in unique control problems. Finally, because much of the material in the later chapters involves model-based control, it is important to understand the techniques of mathematical modeling, the basis of such models of polymerization systems, and their relative merits. Certainly a full course in polymerization reaction engineering would be beneficial to the person wishing to work in polymerization control, but this text will introduce that subject when a dedicated course is not available.

With the material presented in the first four chapters, one can design a polymerization reactor and specify operating conditions. To operate the reactor, however, an understanding of control concepts would be required. The rest of the book is devoted to the study of control concepts relating to polymerization reactors. It is assumed that the reader has had a course in basic control systems concepts and is familiar with sampled data control concepts.

In a polymerization reactor, raw materials are mixed at specified operating conditions to produce polymer(s) having desired properties. The end-use properties of interest include color, viscoelasticity, thermal properties, and mechanical properties among others. To produce a polymer with such desired properties means that process variables such as temperature, molecular weight, molecular weight distribution, and Mooney viscosity must be tightly controlled. The manipulated variables available for controlling the variables of interest at setpoints include the flow rates of raw materials and catalysts, temperature of feed streams and temperature, and/or flow rates of heating/cooling mediums.

Automatic control of polymerization reactors is complicated for the following reasons:

Polymerization processes are highly nonlinear; the use of linear controllers often results in poor performance.

Many polymerization systems are open-loop unstable; therefore safety considerations are of paramount importance.

Polymerization reactor control systems are multivariable in nature. Process interactions, deadtime, and constraints complicate the control systems design of these units.

Many of the important variables such as molecular weight, molecular weight distribution, and Mooney viscosity cannot be measured directly. They must be inferred from other measurements. Inferential measurements can lead to erroneous results and, therefore, automatic control systems that are based on such measurements must be designed to accommodate such errors.

We describe the methods that are available to tackle these and related problems.

In Chapter 5, we describe the polymerization reactor control problem and identify the important end-use properties, controlled variables, manipulated variables, and disturbance variables. Chapter 6 describes the methods for the measurement of important process variables. Methods for estimating those variables that cannot be measured directly are also covered. Next, we describe methods for obtaining a dynamic model of the polymerization reactor. In Chapter 8 we show that in some cases single-input single-output (SISO) strategies can be effectively used for control. Included in this chapter are a number of modern approaches for SISO control. Chapter 9 is concerned with multivariable control strategies for polymerization reactors. Multivariable control must be used where SISO systems and multiloop control do not give adequate performance. In Chapter 10, we describe the nonlinear control techniques that may be applied to polymerization reactors. Chapter 11 is an introduction to polymer processing and to how product properties determined during the polymerization may affect processing. Throughout the second part of the text, the concepts are reinforced by examples and applications.

The authors thank the Georgia Institute of Technology, the University of Louisville, and du Pont, for their support in this endeavor.

F. Joseph Schork
Pradeep B. Deshpande
Kenneth W. Leffew

Contents

Control
of
Polymerization
Reactors

1

Introduction to Polymerization

In this chapter the fundamentals of polymerization as they apply to control of polymerization reactors will be highlighted. Section 1.1 will give a very brief overview of polymeric structure. Sections 1.2 through 1.5 will present the four basic mechanisms of polymerization; a fundamental knowledge of polymer science will be presumed. For a good treatment of the subject, the reader may refer to the texts by Billmeyer [1], Rodriguez [2], Rudin [3], Williams [4] and Odian [5]. Structure–property relationships are well covered by Seymour and Carraher [6].

1.1 Macromolecules

This section will provide a starting point for the discussion of polymerization reactions and their control. The section is broken into three major topics: a brief description of polymer structures, a systematic definition of the molecular weight distribution (MWD), and a brief description of techniques of polymer analysis. An overview of polymer structures is included to motivate later discussions of control of polymer structure. The description of MWD serves to define the nomenclature for later use. The discussion of polymer analysis is meant to lead into later applications of on-line measurements of polymer properties. Recent advances in polymer characterization can be found in Ref. [7].

1.1.1 Structure

The concept of macromolecules is not an old one. Some polymeric products were commercially available in the late 1800s, mostly modified cellulose and other modified natural macromolecules, but an understanding of the nature of macromolecules paralleled the development of the first synthetic polymers in the early 1900s. A polymer is a large molecule made up of repeating units referred to as monomer units. Whereas each monomer unit may have a molecular weight of less than 100, when a polymerization reaction is carried out to join these monomer units, the molecular weight of the resulting polymer can be very high. Most commercial polymers have molecular weights between 10,000 and several million. Specialty polymers and biopolymers such as proteins may have molecular weights that are substantially higher. Examples of common monomers and their resulting polymers are given in Table 1.1.

The synthetic polymer industry began when H. L. Baekeland developed a polymer (Bakelite) resulting from the reaction of phenol and formaldehyde. Formaldehyde reacts with phenol under mild conditions to form a methylol derivative. This, then, forms the repeat unit and will take part in a condensation reaction, under acid conditions with excess phenol ("novolac" resins) or under alkaline conditions with excess formaldehyde ("resole" resins), to form polymers of molecular weights as high as 1000. This corresponds to about 10 repeat units per chain. The number of repeat units per chain is used as a specification of the length of the polymer chain, and is called the *degree of polymerization* $(\bar{x})$. The molecular weight of a given chain is then the degree of polymerization times the molecular weight of the repeat unit. Thus, for the Bakelite example, the molecular weight of the repeat unit is 106. If the degree of polymerization is 10, then the molecular weight of the polymer chain is $10 \times 106 = 1060$.

Much higher molecular weights are possible. For instance, if methyl-methacrylate (monomer) is polymerized in an emulsion system (Chapter 4), molecular weights of several million result. For a polymer (polymethyl-methacrylate) molecular weight of 2,000,000 and a monomer (methylmethacrylate) molecular weight of 100, the average degree of polymerization must be 20,000. That is, there are 20,000 monomer units per polymer molecule. This example also illustrates a different mechanism of polymerization. As Table 1.1 indicates, the polymerization of methylmethacrylate occurs, not by condensation, but by the opening of a double bond to form the polymer backbone.

Most polymerizations take place by one of two general mechanisms, *step-growth or addition polymerization*. Step-growth takes place when functional groups on the monomer molecules react to form dimers or units containing two monomer molecules. This reaction is often a condensation reaction in which water or other low-molecular-weight by-products (HCl, CH_3COOH, etc.) are formed, as in the reaction of phenol with formaldehyde to form a phenolic resin with water as a

Table 1.1 Examples of common high polymers.

Polymer	Monomer	Repeat Unit
Polyethylene	$CH_2{=}CH_2$	$-CH_2CH_2-$
Polymethyl methacrylate	$\underset{}{CH_2{=}\overset{CH_3}{\underset{\mid}{C}}COOCH_3}$	$-CH_2\overset{CH_3}{\underset{\underset{COOCH_3}{\mid}}{\underset{\mid}{C}}}-$
Polyvinyl acetate	$CH_2{=}CH\overset{O}{\overset{\|}{O}}CH_3$	$-CH_2\overset{O(CO)CH_3}{\overset{\|}{C}}H-$
Polystyrene	$CH_2{=}CH$ ⬡	$-CH_2CH-$ ⬡
Polyisoprene (natural rubber)	$CH_2{=}CH\underset{\underset{CH_3}{\mid}}{C}{=}CH_2$	$-CH_2CH{=}\underset{\underset{CH_3}{\mid}}{C}CH_2-$
Polychloroprene (neoprene rubber)	$CH_2{=}\underset{\underset{Cl}{\mid}}{C}CH{=}CH_2$	$-CH_2\underset{\underset{Cl}{\mid}}{C}{=}CHCH_2-$
Polycaprolactam (nylon 6)	$NH_2(CH_2)_5\overset{}{\underset{\underset{O}{\|}}{C}}OH$	$-NH(CH_2)_5\overset{}{\underset{\underset{O}{\|}}{C}}- \ +H_2O$
Polyhexamethylene adipamide (nylon 66)	$HO\overset{O}{\overset{|}{C}}(CH_2)_4\overset{O}{\overset{|}{C}}OH + $ $NH_2(CH_2)_6 NH_2$	$-\overset{O}{\overset{\|}{C}}(CH_2)_4\overset{O}{\overset{\|}{C}}NH(CH_2)_6 NH-$
Phenol formaldehyde	OH ⬡ $+ CH_2O$	OH ⬡$-CH_2-$ $+ H_2O$

by-product. In the idealized case, the dimer units then react with other dimers to form tetramers. The tetramers then react to form larger molecules. The time to form a high-molecular-weight polymer chain may be of the order of hours. This will be important in the design of reactors for step-growth polymerization.

Addition polymerization takes place via the opening of a double bond on the monomer unit. The simplest example is that of ethylene shown in Table 1.1. The double bonds on adjacent ethylene units are opened to allow the formation of a single carbon–carbon bond between the two ethylene units. This leaves a free radical on the second ethylene unit which causes the addition of another ethylene by the opening of its double bond. The reaction continues until the free radical is terminated in some way. The major aspects in which addition polymerization differs from step-growth polymerization are that the linkages are made by the opening of carbon–carbon double bonds and that the time necessary to form a complete molecule of high molecular weight is of the order of seconds.

Thus far, it has been assumed that the polymer molecules in question are linear. If, however, a monomer unit in a condensation polymer contains a third functional group (in addition to the two necessary for the polymeric backbone), or a monomer unit in an addition polymer contains an unreacted double bond or other reactive sites, it is possible to form branched structures as shown in Fig. 1.1a. The term *branching* implies that the individual polymer molecules are discrete. If branching occurs such that a large portion of the polymer chains are interconnected (Fig. 1.1b), the polymer is described as having a *cross-linked* or *network* structure. A cross-linked polymer is referred to as a gel and is insoluble solvents which are inert to the polymer. The gel may coexist with a soluble fraction of the polymer which is not cross-linked (the sol fraction).

The degree of branching affects polymer properties such as density, melt viscosity, and crystallinity. The degree of cross-linking determines the character of the polymer. Non-cross-linked polymers are thermoplastic. Thermoplastic polymers can be melted and cast, extruded, or injection molded. At higher temperatures, thermoplastic polymers undergo thermal decomposition. Cross-linked polymers are thermosetting. A thermosetting polymer, when heated, undergoes thermal decomposition without first becoming fluid because, in a heavily cross-linked material, the entire polymer matrix consists of a single unit and there is no mechanism for flow induced by heating.

As a polymeric material is cooled, it undergoes a second-order transition as is passes through the *glass-transition temperature*. This temperature corresponds to the almost total cessation of molecular motion on the large scale. Below its glass-transition temperature, a polymer has many of the properties associated with an inorganic glass such as hardness and brittleness. Some linear polymers (particularly those without bulky side groups) undergo an additional transition at a temperature significantly above the glass-transition temperature, known as the *crystalline melting point*. Below the crystalline melting point, the polymer chains

Branching

Requires reactive functional group or additional
double bond   (a)

Network

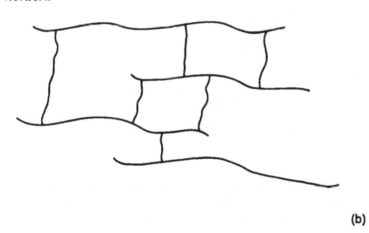

(b)

Figure 1.1 Branching and network formation.

tend to fold back on themselves in a regular fashion. This is illustrated schematically in Fig. 1.2. Not all segments of the polymer need be in a crystalline form.
Microdomains of crystalline material can exist side by side with areas of purely
amorphous material. The fraction of the total polymer which is in the crystalline
state is known as the *degree of crystallinity*.

The geometric arrangement of the atoms in a polymer chain may be classified
as *conformations* and *configurations*. Conformations result from rotations about
single bonds and are transient, whereas configurational changes can take place
only by breaking and making chemical bonds. An example of conformation is

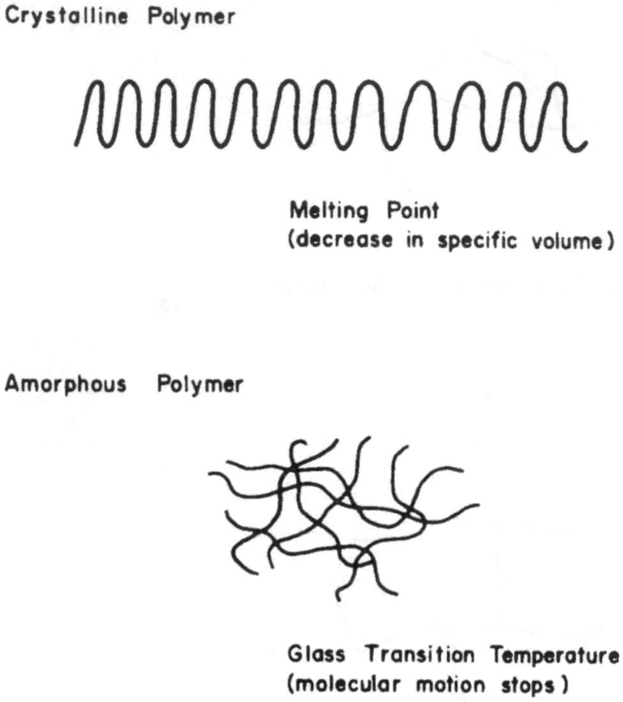

Crystalline Polymer

Melting Point
(decrease in specific volume)

Amorphous   Polymer

Glass Transition Temperature
(molecular motion stops)

Figure 1.2 Crystalline and amorphous polymer.

the multiple orientations that can be taken by a polymer molecule in solution. An example of configuration may be found by considering the placement of side groups off the polymeric backbone as shown in Fig. 1.3. A polymer in which all of the side groups (R) are above or below the plane of the carbon backbone are known as *isotactic*. Configurations in which the side groups alternate regularly above and below the plane of the plane of the carbon backbone are called *syndiotactic*. Finally, configurations in which the side groups occur randomly above or below the carbon backbone are called *atactic*. Tacticity directly affects crystallinity, and hence other polymer properties.

The relationship between structure and properties is critical to the control of polymerization processes since control of the processes will be aimed at manipulating the structure of the polymer in order to develop a polymer with the correct end-use properties. The end-use properties are often evaluated by the use of rather arbitrary tests which attempt to mimic the performance of the polymer in a given end use. An example is the weathering of polymeric coatings under specific environmental conditions. Such end-use properties are not easily correlated

Isotactic

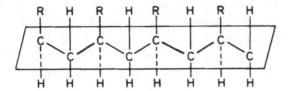

Syndiotactic

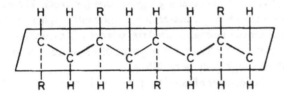

Atactic

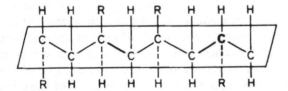

Figure 1.3 Configuration.

with polymeric structure. The discussion that follows attempts to relate polymer structure to more intrinsic properties of the material. The relationship between intrinsic properties and end-use properties must be done for each specific application, and is best left to the formulation chemist.

Thermoplastic Polymers

The nature of a thermoplastic polymer is most significantly affected by three structural factors: degree of crystallinity, molecular orientation, and molecular weight. The degree of crystallinity determines the primary nature of the polymer and, hence, its applications. Highly crystalline materials are hard and brittle. Polymers of medium crystallinity are tough. Low-crystallinity thermoplastics often have a rubbery nature. It should be noted that crystallinity is a direct consequence of molecular structure. Linear polymers with small side groups are highly crystalline. Linear polymers with bulky but regular side groups may crystallize if the side groups fall into some regular folding pattern of the polymer chain. Polymers with bulky but irregular side groups do not crystallize. This last fact can be exploited to produce an amorphous material by randomly including a bulky comonomer into the backbone of the polymer. The lack of regularity prevents crystallization.

The orientation of a linear polymer can greatly change its mechanical properties. The term orientation means the alignment of the linear molecular chains such that their long axes are essentially parallel. This is accomplished in a drawing process in which the polymer is extended along a given axis (or axes). The orientation produced by this process gives the material a significantly higher modulus of elasticity (stiffness) and tensile strength. Uniaxial drawing is exploited in the drawing of fibers in the textile industry to afford yarns of high strength. Biaxial drawing is used to produce tough polymeric films.

The molecular weight of a polymer affects its properties in many ways. When discussing molecular weight, it should be remembered that one is actually discussing the molecular weight distribution. However, when the term molecular weight is used in a general sense, it may be taken to mean some average (number, weight, etc.) molecular weight. A high-molecular-weight material exhibits a higher melt viscosity (above the crystalline melting point) and a higher solution viscosity at constant weight fraction polymer. These properties may be critical in processing steps such as extrusion, injection molding, and wet or melt spinning of fibers. In addition, a higher-molecular-weight material has a higher modulus of elasticity and thus, may, be used in applications where mechanical strength is important.

The breadth of the molecular weight distribution of a polymer may significantly affect its suitability for certain applications. For example, in the extrusion of polypropylene, the best balance of processing characteristics and product tensile strength is achieved when the molecular weight distribution is as narrow as possible [1]. On the other hand, in adhesives formulation, a broad distribution is desirable because the low-molecular-weight material provides adhesion, whereas the high-molecular-weight fraction provides cohesion.

All thermoplastics are viscoelastic. This means that they exhibit properties of both viscous liquids (flow) and elastic solids (elimination of a deformation after release of stress). As with all viscoelastic materials, the stress–strain behavior is time dependent. The classic example of this behavior is the siloxane polymers, sold as toys, which bounce if dropped on the floor, but flow into a puddle if left on the countertop. The relaxation time, or time necessary for relaxation of externally applied stress (flow into a puddle), increases with increases in the molecular weight.

Thermosetting Polymers

The properties of a thermosetting polymer are determined primarily by the nature of the polymer and by the degree of cross-linking. Lightly cross-linked polymers exhibit elastomeric properties. The cross-links provide the toughness and resiliency characteristics of elastomers (rubbers). Highly cross-linked materials are rigid and quite hard. Because they do not flow when heated, thermosetting materials must be polymerized in their final form. This is done by polymerization in a mold. Some polymers, particularly elastomers, can be produced as linear polymers and then cross-linked during the molding operation. Thus, most rubbers are thermoplastic and can be injection molded or extruded. The cross-linking takes place in a "curing" operation after the rubber is in its final form. This may take place in the mold or in a separate processing step.

1.1.2 Molecular Weight Distribution

Thus far, the degree of polymerization for a single polymer molecule has been defined. But not all polymer molecules within a reactor have the same degree of polymerization. Rather, a polymer produced in a single reaction exhibits a distribution of chain lengths (degree of polymerization) as shown in Fig. 1.4. The distribution of chain lengths within a polymeric material may well be the most important factor in determining its end-use properties. Therefore, it will be necessary to develop a method of describing the distribution of chain lengths in a polymeric material.

Following the development of Ray [8], Fig. 1.4a shows the chain length distribution where P_n is the concentration of polymer of length n. The mean of this distribution is the number average chain length (NACL). The same data can be plotted as in Fig. 1.4b where the weight of polymer of a given chain length is plotted against chain length. This is the weight chain length distribution, and its average is the weight average chain length (WACL). If Figs. 1.4a and 1.4b are replotted with the molecular weight of the chains, rather than the chain length on the abscissa, Figs. 1.4c and 1.4d result. These are, respectively, the number molecular weight distribution and the weight molecular weight distribution. Their averages are, respectively, the number average molecular weight (NAMW) and the weight average molecular weight (WAMW). It should be obvious from Fig. 1.4 that, although they are derived from the same distribution data, the values of NAMW and the WAMW will not be necessarily the same. This is because both are single number attempts to represent an entire distribution. It should be noted that the end-use properties of a polymer are determined by the distribution of molecular sizes which is independent of the average used to characterize it. If P_n (the concentration of chains containing n polymer units) is known for all values of n, the various averages may be calculated as follows:

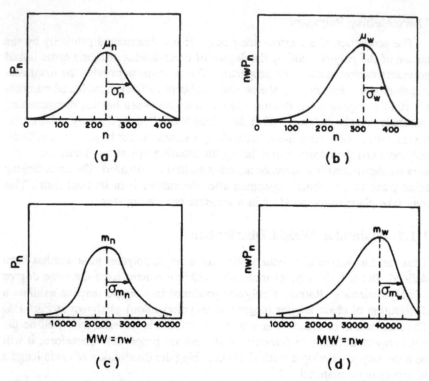

Figure 1.4 Distributions describing polymer chain length. (a) The number chain length distribution (NCLD); (b) the weight chain length distribution (WCLD); (c) the number molecular weight distribution (NMWD); (d) the weight molecular weight distribution (WMWD). (From Ref. 8 with permission.)

$$\text{NACL} \equiv \mu_n \equiv \frac{\sum_{n=1}^{\infty} nP_n}{\sum_{n=1}^{\infty} P_n} \tag{1.1}$$

$$\text{WACL} \equiv \mu_w \equiv \frac{\sum_{n=1}^{\infty} n^2 P_n}{\sum_{n=1}^{\infty} nP_n} \tag{1.2}$$

$$\text{NAMW} \equiv m_n \equiv \frac{\displaystyle\sum_{n=1}^{\infty} (nw)P_n}{\displaystyle\sum_{n=1}^{\infty} P_n} = w\mu_n \tag{1.3}$$

$$\text{WAMW} \equiv m_w \equiv \frac{\displaystyle\sum_{n=1}^{\infty} (nW)^2 P_n}{\displaystyle\sum_{n=1}^{\infty} (nw)P_n} = w\mu_w \tag{1.4}$$

Each of these distributions has a variance associated with it. However, common practice is to judge the breadth of the molecular weight distribution by defining the *polydispersity*, D, given by

$$D \equiv \frac{m_w}{m_n} = \frac{\mu_m}{\mu_n} \tag{1.5}$$

Inspection of Eqs. (1.1) through (1.5) reveals that the polydispersity takes on a value of 1 for a monodisperse sample (one in which all of the chains are of the exact same length). For any other distribution of chain lengths, the polydispersity will be greater than 1. Values of 2 to 5 are common; polydispersities of 100 are known. The disadvantage of using polydispersity to characterize the breadth of the distribution is that small changes in polydispersity can indicate large changes in the chain length distribution. On the other hand, the polydispersity varies only slightly with average chain length and, thus, can be used to compare the breadth of distributions of different average chain lengths.

Variations in degree of polymerization (and hence in molecular weight) occur for at least three reasons. The main mechanism by which the molecular weight distribution is broadened is through the nature of the series-parallel reaction mechanisms leading to chain formation. This effect will be quantified later. A second mechanism is that of spatial or temporal variations in reaction conditions during polymerization. Variations in temperature, monomer concentration, etc., in any reactor, and in residence time in a continuous reactor, affect the individual chain lengths. The final mechanism of variation in degree of polymerization is that of stochastic variations reaction rates on a molecular level. This, however, has been shown to be insignificant in relation to the previous two. The important concept, then, is that a distribution of chain lengths will result due to the nature of the reaction mechanisms, even when all environmental variables (temperature, monomer concentration, etc.) are kept constant.

1.1.3 Methods of Analysis

The analysis of the structure of polymers and of their resulting properties will
be discussed under four major headings: molecular weight, calorimetry, optical
properties, and mechanical properties. Molecular weight and calorimetric
measurements result in information primarily about the structure of the polymer
in question and, indirectly, about its properties. Measurements of optical and
mechanical properties can serve as measures of the suitability of the polymer for
a specific end use, but can also provide information about structure. It should
be evident by now that the differentiation between the structure and the properties of a polymer is not without some ambiguity.

Molecular Weight

The determination of the molecular weight distribution, or of some average
of this distribution, can be accomplished in many ways. Three of the most common will be discussed here: light scattering, intrinsic viscosity, and gel permeation chromatography. In dilute solution, polymer molecules will exist as loose
coils. These conformations are approximately spherical and will scatter incident
light in much the same way as hard spheres. This phenomenon can be exploited
to give absolute measurements of weight average molecular weight. Details of
the technique can be found in Billmeyer [1].

Measurements of intrinsic viscosity may also be used to determine average
molecular weight. If the viscosity of a polymer solution is denoted by η, the intrinsic viscosity of the solution may be defined as

$$\eta_{sp} = \frac{\eta - \eta_o}{\eta_o} \tag{1.6}$$

where η_o is the viscosity of the solvent. The reduced viscosity, η_r, may then be
defined as

$$\eta_r = \frac{\eta_{sp}}{c} \tag{1.7}$$

where c is the polymer concentration in the solvent. The intrinsic viscosity, $[\eta]$,
can now be defined as

$$[\eta] = \lim_{c \to o} \eta_r \tag{1.8}$$

The viscosity average molecular weight, m_v, can then be calculated from the
Mark–Houwink equation,

$$[\eta] = K' (m_v)^a \tag{1.9}$$

where K' and a are calibration constants for the polymer–solvent system in question. The viscosity average molecular weight is be defined as

$$m_v = \left[\frac{\displaystyle\sum_{n=1}^{\infty} (nw)^{(1+a)} P_n}{\displaystyle\sum_{n=1}^{\infty} (nw) P_n} \right]^{1/a}$$

(1.10)

For $a = 1$, $m_v = m_w$. If a polymer is fractionated by extraction into solvent mixtures of increasing solubility, the entire molecular weight distribution can be obtained by intrinsic viscosity analysis on each fraction because for each fraction, $m_n \approx m_v \approx m_w$, absolute molecular weight measurements are possible.

Differences in the size of polymer molecules in dilute solution may be used to fractionate a polymer sample by molecular weight. One specific technique to exploit this possibility is known as gel permeation chromatography (GPC). In this technique, a dilute solution of polymer in an effective solvent (for nonolefinic polymers, tetrahydrofuran (THF) is often used) is injected onto a chromatographic column of porous silica or styrene–divinylbenzene beads. The polymer is then eluted off of the column with additional solvent. The solvated polymer molecules are essentially spherical. A molecule of a specific size will be only able to penetrate pores in the packing which are significantly larger than the diameter of the polymer molecule. If there is a distribution of pore sizes, the effective volume of the column and, thus, the retention time will vary with molecular size. Thus, the largest (highest-molecular-weight) material will be denied access to most of the pore volume and will elute from the column first. Very small molecules will have access to the entire pore volume and will elute last. If the concentration of a polymer in the eluent stream is monitored as a function of eluent volume (by means of UV absorption, dielectric constant, refractive index, or other detector), and the molecular size (weight) is related to the total eluent volume (or retention time) via a calibration curve, the molecular weight distribution can be determined.

By means of multiple columns and slow elution, combined with accurate calibration from known molecular weight standards, it is possible to determine the molecular weight distribution quite accurately. Corrections for axial dispersion during elution may be made if required. Because the detector is measuring polymer concentration (and not number of polymer chains) in the eluent stream, the gel permeation chromatograph yields a weight molecular weight distribution. Gel permeation chromatographs of varying sophistication are commercially available. The most elaborate models feature microprocessor control of the chromatographic process, including automatic sample preparation and multiple sample injection, as well as automatic preparation of the molecular weight distribution from the raw data.

Differential Scanning Calorimetry

A great deal of information about the glass-transition temperature, crystalline melting point, degree of crystallinity, and reactions (degradation, cross-linking, etc.) of a polymer may be determined by thermal analysis. Whereas this may be accomplished by classical adiabatic calorimetry, much simpler instruments for determining the same information have become commonplace over the last 25 years. Of these, the one which yields the most information about the thermal properties of the polymer is the differential scanning calorimeter (DSC). In this technique, small quantities of a sample and a reference material are loaded into identical cells. The reference material (often powdered alumina) should have known specific-heat-versus-temperature characteristics and should exhibit no thermal transformations in the temperature range of interest. The sample and reference cells are then subjected to a ramp increase in temperature (often 10 °C/min). A control circuit apportions power to the sample and reference to keep them at exactly the same temperature. The difference in power inputs is plotted versus temperature.

When the sample undergoes an endothermic transition (such as crystalline melting), the difference in power will be increased because additional power is necessary to provide the heat of melting (approximately the inverse of the heat of crystallization) while maintaining equal temperatures. The heat of crystallization is then the area under the power curve. By knowing the heat of crystallization for a sample of the same polymer with a known degree of crystallinity, it is possible to determine the degree of crystallinity of the sample under analysis.

From the power-versus-temperature curve and a knowledge of the thermal characteristics of the reference, it is possible to calculate the specific heat versus temperature for the sample under analysis. An example of such a curve for quenched (amorphous) polyethylene terephthalate is shown in Fig. 1.5. As the temperature is increased, there is a distinct change in slope as the glass-transition temperature is approached (60–80 °C). With the onset of molecular mobility, crystallization takes place and is indicated by the sharp drop in the specific heat. At a higher temperature (220–270 °C), crystalline melting takes place and is indicated by a sharp rise in specific heat. Thus, the glass-transition temperature, crystalline melting point, and degree of crystallinity can all be determined from a single DSC analysis. In addition, if degradation or cross-linking occurs at high temperatures, the enthalpy of these reactions will be reflected as changes in differential power over the range of temperatures over which these reactions occur.

Mechanical Properties

Determination of mechanical properties most often involves determination of the stress–strain curve of the material in question. This is done by continuously

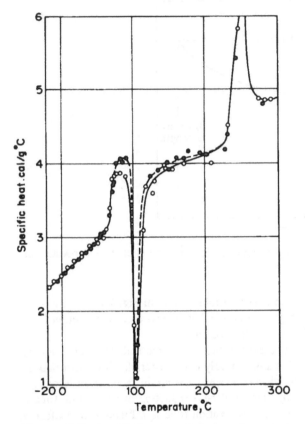

Figure 1.5 Specific heat as a function of temperature for amorphous polyethylene terephthalate. (From Ref. 9 with permission.)

measuring the tensile strength as the sample is elongated at a constant rate of extension (constant rate of strain). Fully automated commercial instrumentation for this procedure is common in polymer laboratories. A typical result for a plastic material is shown in Fig. 1.6. Note that there is a region at low strain in which the developed stress (tensile force divided by cross-sectional area) is linear with strain (percent elongation). The material thus acts as a Hookean spring, and the spring constant (the slope of the stress–strain curve) is known as the *modulus* or *elastic modulus*. Beyond a certain strain (*elongation at yield*), the sample extends rapidly ("necking") with little increase in stress. The stress at which yield occurs is known as the *yield stress*. After additional extension, the sample breaks. This point is denoted by the *ultimate strength* and *elongation at break*. Polymeric

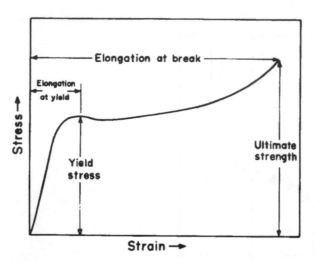

Figure 1.6   Stress–strain curve for a typical polymer. (From Ref. 10 with permission.)

materials will have stress–strain curves similar to that in Fig. 1.6. Highly cross-linked materials will have a high modulus, whereas low molecular weight linear polymers will exhibit yield at very low stress.

Tensile testing is done over a reasonably short time and, thus, does not accurately reflect the viscoelastic nature of polymeric materials. Such materials are elastic at high strain rates and viscous at very low strain rates. If a viscoelastic material is elongated and held, there will be a reduction in stress with time as viscous flow causes relaxation. The relaxation time is a function of molecular weight and temperature. If, on the other hand, the material is stressed, and then the stress is removed, there will be a reduction in strain as the elasticity of the material causes it to attempt to return to its original length. Any viscous flow which occurred during the stressing will not be recovered. Polymer melts and polymer solutions exhibit viscoelastic behavior as well. As might be anticipated, the viscous component is much more significant in these systems. However, viscoelastic effects such as relaxation are definitely present in these systems. The rheology of polymer melts will be treated in Chapter 11.

1.2 Anionic Polymerization

In the rest of this chapter, the four major kinetic mechanisms of polymerization (anionic, cationic, free radical, and step-growth) will be described. A discussion of coordination polymerization will be deferred until Chapter 4 because its mechanism is inherently heterogenous. Addition polymerization can be carried out by a number of mechanisms. The free radical mechanism is commercially predominant, but addition polymerization is often carried out by anionic and cationic mechanisms. Anionic polymerization takes place via the opening of a carbon double bond on the monomer unit. Initiation takes place with the addition of a negative ion to the monomer, resulting in the opening of a double bond and growth at the end bearing the negative charge. Propagation proceeds by addition of monomer units with the carbanion remaining with the propagating chain end. Termination of a growing chain usually involves transfer, and only results in the net loss of a growing chain if the new species is too weak to propagate. Because termination usually involves transfer to some impurity in the system, it is possible, with carefully purified reagents, to carry out polymerization in which termination is lacking. The resulting species are termed *living polymers* and may result in extremely narrow (essentially monodisperse) molecular weight distributions.

Anionic polymerization is employed with vinyl monomers containing electron-withdrawing groups such as nitrile, carboxyl, phenyl, or vinyl in an aprotic non-polar solvent. It is characterized by high rates of polymerization and low polymerization temperatures. Strong bases such as alkyl metal amides, alkoxides, alkyls, hydroxides, and cyanides are often used to form the original carbanion.

$$AC \overset{K}{\Longleftrightarrow} A^- + C^+ \Bigg\}$$ (1.11)

$$\phantom{AC \overset{K}{\Longleftrightarrow}} \text{Initiation}$$

$$A^- + M \overset{k_i}{\Longrightarrow} AM^-$$ (1.12)

$$AM_n^- + M \overset{k_p}{\Longrightarrow} AM_{n+1}^- \qquad \text{Propagation}$$ (1.13)

$$AM_n^- + B \overset{k_f}{\Longrightarrow} AM_n + B^- \qquad \text{Chain Transfer}$$ (1.14)

Here A^- is the anion initiating the polymerization, B is the chain transfer molecule, and B^- is the new anion formed by chain transfer, which may or may not be capable of initiation of a new chain. The mechanism can be written in a form which is more concise and consistent with the subsequent treatment for free radical polymerization as

$$AC \overset{K}{\Longleftrightarrow} A^- + C^+ \Bigg\}$$ (1.15)

$$\phantom{AC \overset{K}{\Longleftrightarrow}} \text{Initiation}$$

$$A^- + M \overset{k_i}{\Longrightarrow} P_1$$ (1.16)

$$P_n + M \overset{k_p}{\Longrightarrow} P_{n+1} \qquad \text{Propagation}$$ (1.17)

$$P_n + B \overset{k_f}{\Longrightarrow} M_n + B^- \qquad \text{Chain Transfer}$$ (1.18)

Here P_n is taken to mean AM_n^-, and M_n represents AM_n.

For very fast reactions, the concentration of reactive species (in this case, ionic chains) becomes essentially constant very early in the reaction. For this to happen here, the rates of initiation and chain transfer must reach steady state quickly and be equal. This is known as the quasi-steady-state approximation (QSSA). Based on the mechanism above and making the QSSA for P_n, the rate of polymerization can be written as

$$R_p = k_p PM = \frac{k_i k_p A^- M^2}{k_f B} = \frac{K k_i k_p (AC) M^2}{k_f C^+ B}, \qquad P = \sum_{n=1}^{\infty} P_n \quad (1.19)$$

The QSSA is only valid for significant chain transfer to an unreactive anion, B^-. In the absence of rapid chain transfer (or of chain transfer to a nonpropagating anion), the rate of polymerization will continue to rise as the total number of living chains increases until initiation is complete. Initiation is complete when all of the catalyst has been consumed; from this point on, the number of live chains will remain constant. The instantaneous degree of polymerization may be written as the rate of propagation divided by the rate of chain transfer (rate of production of dead chains):

$$\bar{x} = \frac{k_p M}{k_f B} \qquad\qquad\qquad\qquad (1.20)$$

In Eqs. (1.19) and (1.20), the species symbols are interpreted as concentrations.

1.3 Cationic Polymerization

Cationic polymerization is another mechanism of addition polymerization. It proceeds through chain propagation via a carbonium ion with the opening of a double bond on the monomer unit as with anionic polymerization. The carbonium ion is formed by the reaction of a strong Lewis acid (catalyst) with a weak Lewis base (cocatalyst) followed by attack on the double-bonded monomer unit. Termination via terminal double-bond formation and chain transfer to monomer and polymer are dominant.

Cationic polymerization is carried out with vinyl monomers containing electron-releasing groups such as alkoxy, phenyl, and vinyl. The system is characterized by very high rates of polymerization. The mechanism of cationic polymerization may be written as

$$A + RH \overset{K}{\rightleftharpoons} H^+ AR^- \qquad\qquad\qquad (1.21)$$
$$\left. \begin{array}{c} \\ \\ \end{array} \right\} \quad \text{Initiation}$$
$$H^+ AR^- + M \overset{k_i}{\longrightarrow} HM^+ AR^- \qquad\qquad (1.22)$$

$$HM_n^+AR^- + M \xrightarrow{k_p} HM_{n+1}^+AR^- \qquad \text{Propagation} \qquad (1.23)$$

$$HM_n^+AR^- \xrightarrow{k_i} M_n + H^+AR^- \qquad \text{Termination} \qquad (1.24)$$

$$HM_n^+AR^- + M \xrightarrow{k_f} M_n + HM^+AR^- \qquad \text{Chain Transfer} \qquad (1.25)$$

Here A is the catalyst and RH is the cocatalyst. These two species react to form the catalyst–cocatalyst complex in Eq. (1.21). This complex donates a proton to the monomer, forming a carbonium ion. Because cationic polymerization is usually carried out in a chlorinated hydrocarbon solvent of low dielectric constant, the anion (AR^-) cannot be separated from the carbonium ion. Rather, the two form an intimate ion pair. Propagation takes place by the addition of monomer to the growing chain end [Eq. 1.23]. Termination occurs with the formation of a terminal double bond and the regeneration of the catalyst–cocatalyst complex [Eq. (2.38)]. Chain transfer to monomer takes place as shown in Eq. (1.24).

The mechanism can be written in a form which is more concise and consistent with the previous notation as

$$A + RH \xrightleftharpoons{K} H^+AR^- \Bigg\}$$
$$\qquad\qquad\qquad\qquad \text{Initiation} \qquad\qquad\qquad (1.26)$$
$$H^+AR^- + M \xrightarrow{k_i} P_1 \Bigg\} \qquad\qquad\qquad\qquad\qquad (1.27)$$

$$P_n + M \xrightarrow{k_p} P_{n+1} \qquad \text{Propagation} \qquad (1.28)$$

$$P_n \xrightarrow{k_t} M_n + H^+AR^- \qquad \text{Termination} \qquad (1.29)$$

$$P_n + M \xrightarrow{k_f} M_n + P_1 \qquad \text{Chain Transfer} \qquad (1.30)$$

Here P_n is taken to mean $HM_n^+AR^-$.

Based on the mechanism above, the rate of polymerization may be written as the product of the propagation rate constant, the monomer concentration, and the concentration of live chains (P). If the quasi-steady-state approximation is made for P_n, the rate of polymerization can be written as

$$R_p = k_p PM = \frac{Kk_i k_p A(AH)M^2}{k_t}, \qquad P = \sum_{n=1}^{\infty} P_n \qquad (1.31)$$

If, as is often the case in ionic polymerization, termination is negligible, the QSSA is not applicable and the term after the second equality in Eq. (1.31) cannot be used. In this case, the rate of polymerization will continue to rise as the total number of living chains increases until initiation is complete. Initiation is complete when all of the catalyst has been consumed; from this point on, the number of live chains will remain constant. The instantaneous degree of polymerization may be written as the ratio of propagation to the sum of the rates of termination and transfer:

$$\bar{x} = \frac{k_p PM}{k_t P + k_f PM} = \frac{k_p M}{k_t + k_f M} \qquad (1.32)$$

Thus, if transfer predominates, the degree of polymerization is a function only of temperature (through k_p/k_f). If termination predominates and if the activation energy for termination is greater than the sum of the activation energies for initiation and propagation (as is often the case), both the rate of polymerization and the degree of polymerization increase with decreasing temperature. These conditions are the reverse of those found in free radical polymerization and allow the attainment of high molecular weight.

1.4 Free Radical Polymerization

Free radical polymerization is the most common of all addition polymerization mechanisms. When free radicals are generated in the presence of unsaturated monomers, the radical adds to the double bond and another radical is generated by the resultant unpaired electron. This radical is free to react with another monomer unit, and in this way the polymer molecule grows by adding monomer units while maintaining a free radical at the reactive end of the live (growing) chain. Chain growth continues until the radical is terminated or transferred to another chain. The complete mechanism can be written as follows:

$$I \xrightarrow{k_d} 2R \qquad \qquad \qquad (1.33)$$
$$\left. \begin{array}{l} \\ \\ \end{array} \right\} \quad \text{Initiation}$$
$$M + R \xrightarrow{k_i} P_1 \qquad \qquad (1.34)$$

$$P_n + M \xrightarrow{k_p} P_{n+1} \qquad \text{Propagation} \qquad (1.35)$$

$$P_n + P_m \xrightarrow{k_{tc}} M_{n+m} \qquad \text{Termination by Combination} \qquad (1.36)$$

$$P_n + P_m \xrightarrow{k_{td}} M_n + M_m \qquad \text{Termination by Disproportionation} \qquad (1.37)$$

$$P_n + M \xrightarrow{k_{fm}} M_n + P_1 \qquad \text{Chain Transfer to Monomer} \qquad (1.38)$$

$$P_n + S \xrightarrow{k_{fs}} M_n + S\cdot \qquad \text{Chain Transfer to Solvent} \qquad (1.39)$$

$$P_n + T \xrightarrow{k_{ft}} M_n + T\cdot \qquad \text{Chain Transfer to Transfer Agent} \qquad (1.40)$$

$$P_n + M_m \overset{kf_p}{\rightleftharpoons} M_n + P_m \qquad \text{Chain Transfer to Polymer} \qquad (1.41)$$

$$R + In \overset{kin}{\rightleftharpoons} Q \qquad\qquad \text{Inhibition} \qquad\qquad (1.42)$$

This complex set of reactions may be divided into initiation, propagation, termination, and chain transfer reactions. As shown in Eq. (1.33), an initiator (I) such as benzoyl peroxide decomposes thermally at the reaction temperature to form two primary radicals (R). Each radical may add to the double bond of a monomer unit [Eq. (1.34)], producing a live polymer chain with a length of one monomer unit (P_1). Live polymer chains (each containing a free radical at its reactive end) propagate by adding monomer as in Eq. (1.35), with a radical always existing at the reactive end of the growing chain. Termination occurs either by combination of two live chains to form a dead chain (M_{n+m}) of length equal to the sum of the lengths of the live chains [Eq. (1.36)] or by disproportionation in which two dead chains $(M_n$ and $M_m)$ are formed, each with a length equal to that of one of the live chains [Eq. (1.37)]. Transfer of the free radical to another polymer chain can occur as shown in Eqs. (1.38)–(1.41). If the radical is transferred to monomer, the original live chain becomes dead, whereas a new chain of length 1 begins to grow. In a similar reaction, the solvent (S) can serve to transfer the radical [Eq. (1.39)]. Likewise, a chain transfer agent (T) can be added to the reactor to enhance radical transfer and, thereby, reduce the molecular weight. The solvent radical $(S\cdot)$ or the transfer agent radical $(T\cdot)$ can then transfer the radical to monomer in reactions analogous to Eq. (1.34). If chain transfer to polymer occurs, a branch begins to form on what was previously a linear dead polymer chain. If an inhibitor (In) such as hydroquinone is present, primary radicals will be converted to an unreactive form (Q) as shown in Eq. (1.42).

Experimental observation of free radical polymerization yields the following general characteristics: early appearance of high-molecular-weight polymer, presence of monomer throughout the polymerization, inhibition of the reaction by additives which react with free radicals without initiation of a polymer chain, and autocatalytic kinetics. These observations result from the chain reaction mechanism. As in ionic polymerization, a high-molecular-weight polymer appears early and a monomer is present late in the reaction because in chain polymerization, unlike step-growth polymerization where each chain grows to its full length in a very short time (order of magnitude of seconds). The progress of the polymerization can then be quantified by the *monomer conversion*, defined as the fraction of the initial monomer which has been converted to a polymer at any given time.

The rate of polymerization may be derived by applying mass-action kinetics to the elementary reactions in Eqs. (1.33)–(1.42). Mass balances for the species R and P (where P is the sum of all P_n) yield

$$\frac{dR}{dt} = 2k_d f I - k_i RM \tag{1.43}$$

$$\frac{dP}{dt} = k_i RM - k_t P^2 \tag{1.44}$$

where f is the initiator efficiency and $k_t = k_{t_c} + k_{t_d}$. The initiator efficiency represents the fraction of the radicals formed by Eq. (1.33) which is successful in initiating chains by Eq. (1.34). Initiator efficiencies generally range from 0.2 to 0.7. There are two major causes for initiator inefficiency [3]: induced decomposition and side reactions. Induced decomposition occurs when a growing polymer chain undergoes chain transfer to the initiator (I). The radical concentration is not increased by this reaction and, hence, the initiator has been "wasted." Induced decomposition is generally negligible for azo initiators, but may be significant for peroxides. The major cause of initiator inefficiency involves side reactions. Immediately after initiator decomposition, the two radicals are trapped in a "cage" of solvent and monomer molecules. During the short interval (approximately 10^{-10} sec.) before one of the two radicals diffuses away, the radicals are far more likely to react with each other than with any other species. These reactions may be recombination to reform the initiator or any other side reactions between the radicals which result in species which cannot initiate polymerization.

Unlike in ionic polymerization, the quasi-steady-state approximation is almost uniformly valid in free radical systems. If the QSSA is made for the radical species R and P (that is, if dR/dt and dP/dt are assumed to be zero), the concentration of live chains may be written as

$$P = \left[\frac{2fk_d I}{k_t} \right]^{1/2} \tag{1.45}$$

Finally, the rate of polymerization, R_p, may be written as

$$R_p = -\frac{dM}{dt} = k_p MP = k_p M \left[\frac{2fk_d I}{k_t} \right]^{1/2} \tag{1.46}$$

In the case where inhibition is significant, Eq. (1.45) must be replaced with [3]

$$P = \left(\frac{2fk_d I}{k_t} \right) \left(\frac{1}{1 + k_{in} In/k_i M} \right)^{1/2} \tag{1.47}$$

Because the inhibition reaction occurs at a much more rapid rate than the initiation reaction, the presence of an inhibitor in a batch reactor will result in an induction phase at the beginning of the reaction, during which no polymerization takes place.

The autocatalytic nature of free radical polymerization comes about due to the gel or Trommsdorf effect [12]. This is a viscosity-induced phenomenon caused by the reduced diffusivity of macromolecules at high conversion (high viscosity). Because termination is a bimolecular reaction involving macromolecules, it is affected most severely. The termination rate falls due to diffusional resistance to reaction. This is correlated as a reduction in k_t. Fig. 1.7 shows the large changes in the ratio of k_t to k_{t0} (the value of k_t at zero conversion) as monomer conversion increases. As the rate of termination falls, the rates of the other reactions are essentially unaffected, as they are at most monomolecular in macromolecules. Thus, as the rate of termination falls, the rate of polymerization increases as shown in Eq. (1.46). The reaction is autocatalytic because the rate of reaction increases with accumulation of product (polymer). A similar viscosity effect (glass effect) retards the propagation reaction. However, because propagation involves the reaction of a small monomer molecule with a large chain, the effect is only seen at much higher viscosity (higher conversion). It can be seen, however, when vinyl monomers are polymerized in bulk substantially below the glass-transition temperature of the polymer. In this case, the polymerization will not go to complete conversion if the reaction mixture reaches the glass-transition temperature of the monomer/polymer solution. Older treatments of the glass and gel effects have incorporated these effects into the rate constants for propagation and termination. That is, the rate constants have been made empirical functions of monomer conversion and temperature as above. More recent studies [14–22] have attempted to develop mechanistic models of these effects based on free volume theory. The result is that propagation and termination become chain length dependent.

The instantaneous degree of polymerization can be defined as the rate of propagation divided by the rate of production of dead chains (the sum of the rates of all reactions leading to dead chains):

$$\bar{x} = \frac{k_p M P}{\frac{1}{2} k_{tc} P^2 + k_{td} P^2 + k_{fm} M P + k_{fs} P S + k_{ft} P T} \tag{1.48}$$

Thus, if the rate of polymerization is increased by increasing the initiation rate, the degree of polymerization will fall. A reduction in k_t caused by the gel effect will result in an increase in the degree of polymerization. At very high conversion in high viscosity systems, the degree of polymerization may fall again as the propagation reaction becomes viscosity-limited. If it is desirable to produce lower-molecular-weight material for a given application, a chain transfer agent may be added to reduce the average live chain length as shown in Eq. (1.40). Equation (1.48) shows the inverse relationship between added chain transfer agent (T) and degree of polymerization.

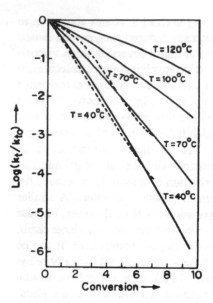

Figure 1.7 The gel or Trommsdorf effect for methyl methacrylate. (From Ref. 13 with permission.)

Chain branching occurs via chain transfer to a polymer which becomes significant at high monomer conversion and low solvent concentration. It is particularly important in bulk and emulsion polymerization systems (Chapter 4) where there is no solvent present to dilute the polymer from attack by free radicals. Branching will reduce the product crystallinity as discussed previously. If necessary, branching may be controlled by stopping the polymerization short of full conversion (perhaps 85–90%) and recovering and recycling the residual monomer.

Cross-linking occurs in monomers with more than one double bond per molecule. If the reactivities of the double bonds are significantly different, it is possible to produce a linear (or branched) polymer by opening one of the double bonds, which can then be cross-linked in a separate reaction (such as vulcanization) in which the second double bond is opened to form the cross-links. It is also possible to induce cross-linking through additional functional groups on the monomer unit. Undesired cross-linking will produce an insoluble gel which may interfere with subsequent processing.

1.5 Step-Growth Polymerization

Step-growth polymerization involves reaction of functional groups on adjacent monomer molecules with the evolution of water or other low-molecular-weight by-products. The condensation of phenol and formaldehyde to form phenol formaldehyde resins has been discussed already and is shown in Table 1.1. Another example is the condensation of caprolactam with itself to form polycaprolactam (nylon 6) and water, also shown in Table 1.1. The reaction is stepwise or step-growth in the sense that the reaction of each functional group is essentially independent of previous condensation reations. There are no activated species as in addition polymerization.

Condensation polymerizations are of two general types. The first, known as A-B condensation is typified by the polymerization of polycaprolactam (nylon 6). The amine group on one caprolactam molecule reacts with the acid group of another caprolactam molecule, joining the two molecules and producing water as a by-product. Because both of the functional groups taking part in the condensation occur on a single type of monomer, the reaction is said to be of the A-B type. If the two functional groups involved in the condensation are on different types of molecules, the polymerization is said to be of the A-A/B-B type. This kinetic scheme is typified by the reaction of hexamethylene diamine with adipic acid to form polyhexamethylene adipimide (nylon 66) as shown in Table 1.1. In this case one of the amine groups on a hexamethylene diamine molecule reacts with one of the acid groups on an adipic acid molecule with the condensation of water. The reaction continues to form high polymers. Because the diamine contains two amine groups, whereas the acid contains two acid groups, alternating structure is formed. The kinetics are represented as A-A/B-B where the A-A represents the diamine and the B-B represents the diacid.

Experimental observation of step-growth polymerization yields the following general characteristics: early disappearance of the monomer, absence of any high polymer during the early stages of reaction, and equilibrium between polymerization and depolymerization reactions. These observations suggest a mechanism of linear condensation in which monomer molecules react to form dimers, the dimers react with each other to form tetramers (or with other oligomers to form larger oligomers), and the tetramers react with other oligomers to form longer chains. Thus, in this step-growth mechanism, the monomer disappears rapidly as it is converted to a dimer. A great deal of low-molecular-weight material is formed early in the reaction, and the average chain length grows slowly as the polymer chains condense to form longer chains. Because the condensation reaction is reversible, the polymerization is always in equilibrium with the depolymerization reaction (hydrolysis). The depolymerization can be controlled by continuously removing the water (or other by-product) of condensation, thus driving the polymerization to completion. The validity of this sort of polymerization mechanism has been verified for a large number of linear condensation polymerizations.

Rate expressions for step-growth polymerization can be written from mass-action kinetics once the mechanism is understood [1]. The condensation is catalyzed by acids. Thus, for a driven system, the rate of polymerization in the presence of an acid can be written as

$$R_p = -\frac{dA}{dt} = kABH^+ \tag{1.49}$$

where A and B are taken to be the concentrations of functional groups A and B. For a stoichiometric ratio of A and B, and assuming the acid concentration to be constant over the reaction, the rate of polymerization may be simplified to

$$R_p = -\frac{dA}{dt} = k'A^2 \tag{1.50}$$

In the absence of added strong acid, an acid functional group on the monomer can catalyze the reaction. The kinetics then become

$$R_p = -\frac{dA}{dt} = k''A^2B \tag{1.51}$$

where A now represents the acidic functional group. For a stoichiometric ratio of functional groups, this becomes

$$R_p = -\frac{dA}{dt} = k'''A^3 \tag{1.52}$$

The progress of the polymerization reaction can be quantified by introducing the *extent of reaction*, p, defined as the fraction of A or B functional groups which has reacted at time t. The number average chain length is given by the total number of monomer molecules initially present divided by the total number of molecules present at time t, which can be related to the extent of reaction as follows:

$$\mu_n = \frac{N_0}{N} = \frac{N_0}{N_0(1-p)} = \frac{1}{1-p} \tag{1.53}$$

Inspection of Eq. (1.53) will indicate that to obtain the necessary high number average chain length, the extent of reaction must be well above 0.99. This is one of the most significant characteristics of step-growth polymerization and will affect significantly the design of reactors for such polymerizations. As noted previously, to prevent depolymerization, it is necessary to remove the by-product to drive the polymerization to a high extent of reaction. This is absolutely necessary in light of the dependence of molecular weight on the extent of the reaction.

In a polymerization of the $A-A/B-B$ type, an imbalance in the stoichiometric ratio of $A-A$ to $B-B$ will result in a marked reduction in molecular weight. If the stoichiometric ratio (r) is defined as the ratio of number of A and B functional groups originally present (always defined to be less than unity),

$$r \equiv \frac{N_A}{N_B} \tag{1.54}$$

the number average molecular weight possible at a given value of r may be determined as follows. The total number of monomer molecules is given by $(N_A + N_B)/2$ or $N_A(1 + 1/r)/2$. The extent of reaction is defined as the fraction of A groups which have reacted at a given time. The fraction of B groups which have reacted is given by rp. The total numbers of unreacted A and B groups are $N_A(1 - p)$ and $N_B(1 - rp)$, respectively. The total number of polymer chain ends is the total number of unreacted A and B groups. The total number of polymer molecules is one-half the total number of chain ends or $[N_A(1 - p) + N_B(1 - rp)]/2$. The number average chain length is the total number of monomer molecules originally present divided by the total number of polymer molecules:

$$\mu_n = \frac{N_A(1 + 1/r)/2}{[N_A(1 - p) + N_B(1 - rp)]/2} = \frac{1 + r}{1 + r - 2p} \tag{1.55}$$

This effect is illustrated in Fig. 1.8. An intentional imbalance in the stoichiometric ratio can be used to control the molecular weight in an A–A/B–B type polymerization. On the other hand, an unintentional stoichiometric imbalance will result in polymer of an undesirable low molecular weight. It should be noted that only slight changes in r can have profound effects on the quality of the product.

In an A–B type polymerization, molecular weight control may be exercised by the addition of a small amount of monofunctional monomer. Thus, the chains are *capped* with the monofunctional monomer, and no more polymerization occurs. If r is redefined as

$$r = \frac{N_A}{N_B + 2N_{B'}} \tag{1.56}$$

where $N_{B'}$ is the number of monofunctional monomer molecules present initially, Fig. 1.8 may be applied to the case of capping with monofunctional monomer. Once again, the stoichiometry is critical to the molecular weight development.

If monomers with more than two functional groups per molecule are used, three-dimensional polymer structures will be formed. Small quantities of trifunctional (or greater) monomers will result in a branched structure as discussed previously. The presence of large quantities of monomer of functionality three or greater will result in the formation of essentially infinitely large polymer networks. This is known as gelation or cross-linking. The onset of gelation results in the division of the polymerizing material into two fractions: an insoluble *gel* of cross-linked material and a soluble *sol* of essentially linear or branched molecules. As with other aspects of the step-growth kinetics, the addition of polyfunctional monomers must be very tightly controlled to produce a desired product.

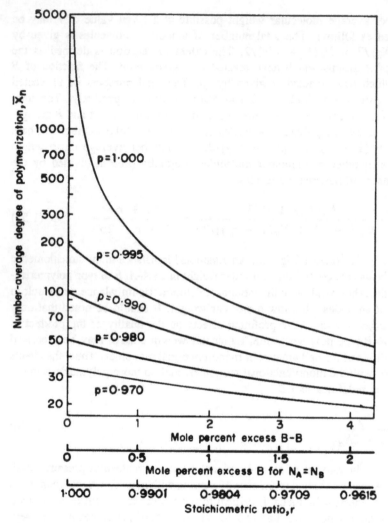

Figure 1.8 Dependence of number average chain length on stoichiometric ratio. (From Ref. 5 with permission.)

References

1. F. W. Billmeyer, Jr., *Textbook of Polymer Science*, 3rd ed., Wiley, New York, 1984.
2. F. Rodriguez, *Principles of Polymer System*, 2nd ed., McGraw-Hill, New York, 1982.
3. Alfred Rudin, *The Elements of Polymer Science and Engineering*, Academic Press, New York, 1982.

4. David J. Williams, *Polymer Science and Engineering*, Prentice-Hall, Englewood Cliffs, NJ, 1971.
5. G. Odian, *Principles of Polymerization*, 2nd ed., Wiley, New York, 1981.
6. Raymond B. Seymour and Charles E. Carraher, *Structure-Property Relationships in Polymers*, Plenum Press, New York, 1984.
7. H. G. Barth and J. Janca (eds.), *Polymer Analysis and Characterization, III*, Applied Polymer Symposium 48, Wiley-Interscience, New York, 1990.
8. W. H. Ray, "On the Mathematical Modeling of Polymerization Reactors," *J. Macromol. Sci. – Revs. Macromol. Chem.*, C8, 1–56 (1972).
9. C. W. Smith and M. Dole, "Specific Heat of Synthetic High Polymers. VII. Polyethylene Terephthalate," *J. Polymer Sci.*, 20, 37–56 (1956).
10. Charles C. Winding and Gordon D. Hiatt, *Polymeric Materials*, McGraw-Hill, New York, 1961.
11. W. E. Houston and F. J. Schork, "Adaptive Predictive Control of a Semibatch Polymerization Reactor," *Polymer Process Engineering*, 5, No. 1, 119–147 (1987).
12. E. Trommsdorf, H. Kohle, and P. Lagally, "On the Polymerization of Methylmethacrylate," *Makromol. Chem.*, 1, 169–198 (1948).
13. R. Jaisinghani and W. H. Ray, "On the Dynamic Behaviour of a Class of Homogeneous Continuous Stirred Tank Polymerization Reactors," *Chem. Eng. Sci.*, 32, 811–825 (1977).
14. S. K. Soh and D. C. Sundberg, "Diffusion-Controlled Vinyl Polymerization. I. The Gel Effect," *J. Polym. Sci. Chem.*, 20, 1299–1313 (1982).
15. S. K. Soh and D. C. Sundberg, "Diffusion-Controlled Vinyl Polymerization. II. Limitations on The Gel Effect," *J. Polym. Sci. Chem.*, 20, 1315–1329 (1982).
16. S. K. Soh and D. C. Sundberg, "Diffusion-Controlled Vinyl Polymerization. III. Free Volume Parameters and Diffusion-Controlled Propagation," *J. Polym. Sci. Chem.*, 20, 1331–1344 (1982).
17. S. K. Soh and D. C. Sundberg, "Diffusion-Controlled Vinyl Polymerization. IV. Comparison of Theory and Experiment," *J. Polym. Sci. Chem.*, 20, 1331–1344 (1982).
18. Thomas J. Tulig and Matthew Tirrell, "Toward a Molecular Theory of the Trommsdorf Effect," *Macromolecules*, 14, 1501–1511 (1981).
19. Dennis J. Coyle, Thomas J. Tulig, and Matthew Tirrell, "Finite Element Analysis of High Conversion Free-Radical Polymerization," *I. &E. C. Fund.*, 24, 343–351 (1985).
20. D. Bhattacharya and A. E. Hamielec, "Bulk Thermal Copolymerization of Styrene/p-Methylstyrene: Modelling Diffusion-Controlled Termination and Propagation using Free-Volume Theory," *Polymer*, 27(4), 611–618 (1986).
21. K. M. Jones, D. Bhattacharya, J. L. Brash, and A. E. Hamielec, "Investigation of the Kinetics of Copolymerization of Methyl Methacrylate/p-Methyl Styrene to High Conversion: Modelling Diffusion-Controlled Termination and Propagation by Free-Volume Theory," *Polymer*, 27(4), 602–610 (1986).
22. S. Zhu and A. E. Hamielec, "Chain Length Dependent Termination for Nonisothermal Free Radical Polymerization," *Polymeric Materials Science and Engineering, Proceedings of the ACS Division of Polymeric Materials Science and Engineering*, 58, American Chemical Society, Washington, D.C., 1988, pp. 393–399.

Bibliography

J. D. Ferry, *Viscoelastic Properties of Polymers*, Wiley, New York, 1970.

P. J. Flory, *Principles of Polymer Chemistry*, Cornell University Press, Ithaca, NY, 1953.

J. P. Kennedy, *Cationic Polymerization of Olefins*, Wiley-Interscience, New York, 1975.

G. R. Moore and D. E. Kline, *Properties and Processing of Polymers for Engineers*, Prentice-Hall, Englewood Cliffs, NJ, 1984.

Herbert Morawetz, *Polymers: The Origin and Growth of a Science*, Wiley-Interscience, New York, 1985.

Maurice Morton, "Anionic Polymerization," in *Kinetics and Mechanisms of Polymerization, Volume 1*: Vinyl Polymerization (Part II) (G. E. Ham, ed.), Marcel Dekker, Inc., New York, 1969.

E. B. Nauman, "Polymer Reaction Engineering," in *Chemical Reactor Design*, Wiley, New York, 1987.

Leighton H. Peebles, Jr., *Molecular Weight Distributions in Polymers*, Interscience, New York, 1971.

Jan F. Rabek, *Experimental Methods in Polymer Science*, Wiley, New York, 1980.

R. B. Seymour and C. E. Carraher, *Polymer Chemistry: An Introduction*, Marcel Dekker, Inc., New York, 1981.

Robert J. Samuels, *Structured Polymer Properties*, Wiley-Interscience, New York, 1974.

M. Szwarc, *Carbanions, Living Polymers, and Electron Transfer Processes*, Wiley-Interscience, New York, 1968.

Zehev Tadmor and Costas G. Gogos, *Principles of Polymer Processing*, Wiley-Interscience, New York, 1979.

Roy W. Tess and Gary W. Poehlein (eds.), *Applied Polymer Science*, 2nd. ed., ACS Symposium Series No. 285, American Chemical Society, Washington, D.C., 1985.

L. R. G. Treloar, *Introduction to Polymer Science*, Wykeham, London, 1974.

Z. Zlamal, "Mechanisms of Cationic Polymerization," in *Kinetics and Mechanisms of Polymerization, Volume 1: Vinyl Polymerization (Part II)* (G. E. Ham, ed.), Marcel Dekker, Inc., New York, 1969.

2

Kinetic Analysis of Polymerization

In this chapter, the focus will shift from the chemistry of polymerization to the mathematical description and analysis of the polymerization process. In Section 2.1, mass and energy balances will be used to combine the kinetics of the polymerization reactions with the mass and energy flows specific to the reactor type. Various mathematical techniques for characterizing the molecular weight distribution will be presented. In Section 2.2, rigorous kinetic analyses will be developed for anionic, free radical, and step-growth polymerization. In this section, the analyses will presume a batch reactor. Chapter 3 will extend the analysis to other reactor configurations. Finally, in Section 2.3, the same sorts of techniques will be used to analyze copolymerizations.

2.1 Mathematical Modeling Techniques

2.1.1 Mass and Energy Balances

Polymerization reactors can be modeled using the classical techniques of chemical reactor design. Primary to this approach are mass balances over various chemical species as well as an energy balance. In all cases, the knowledge of the chemical kinetics is used to describe the rates of formation of various species or, in the case of the energy balance, to describe the rate of heat generation via reaction. These terms are combined with flow terms specific to the reactor in question.

In the case of heterogeneous systems, terms describing interphase mass or heat transfer may also be included. Some processes may occur so rapidly that the species involved are assumed to be at equilibrium at all times. The principles of mass and energy balances are best described by means of an example. Consider anionic solution polymerization in a single continuous stirred tank reactor (CSTR). For very rapid initiation and in the absence of chain transfer, the reaction is described by Eq. (1.17). A mass balance over a monomer may be written as

$$V \frac{dM}{dt} = Q_f M_f - QM - Vk_p PM, \qquad M(0) = M_0 \tag{2.1}$$

where V is the reactor volume, Q_f and Q are the inlet and outlet volumetric flow rates, respectively, and P and M are taken to be concentrations as previously defined. If P is assumed constant (rapid initiation and no chain transfer), a mass balance over the live polymer may be written as

$$V \frac{dP}{dt} = Q_f P_f - QP, \qquad P(0) = P_0 \tag{2.2}$$

Both of these equations are in the standard form

Accumulation = Inflow − Outflow + Generation (or Consumption) (2.3)

where the generation term may be negative or zero as above. The energy balance may be written as

$$\rho C_p V \frac{dT}{dt}$$

$$= \rho C_p Q_f T_f - \rho C_p QT + V(-\Delta H_p)k_p PM - UA(T - T_j), \qquad T(0) = T_0 \tag{2.4}$$

where ρ and C_p are, respectively, the density and specific heat of the reaction mixture, ΔH_p is the heat of reaction (propagation), and U and A are the overall heat transfer coefficient and heat transfer area for the cooling jacket. The CSTR assumption of perfect mixing has been made here. In view of the viscosity of polymerizing solutions and the effect of micromixing on molecular weight development, it may be desirable in some instances to incorporate a more complex mixing model for the reactor. More complex flow and mixing models will be considered in Chapter 3.

Models of the form of Eqs. (2.1), (2.2), and (2.4) can be integrated numerically (or occasionally, analytically) from specific initial conditions to determine the transient behavior of the system. For design purposes, it may be acceptable to consider only the steady-state solution. This may be obtained by setting the time derivatives to zero and solving for M, P, and T.

2.1.2 Molecular Weight Distribution

In Chapter 1, the number average chain length was derived for each kinetic scheme either through the stoichiometry (step-growth polymerization) or through the concept of the degree of polymerization (chain or addition polymerization). These developments give no information about the breadth of the number chain length distribution due to the series-parallel nature of the reactions, differences in reactor type, and variations in operating conditions. To model the entire number chain length (or other) distribution, it is necessary to develop more complex descriptions of the molecular species.

The approach is best illustrated with an example [1,2]. Consider batch anionic solution polymerization with very rapid initiation and no chain transfer of termination. Mass balances over monomer and polymer have be developed for CSTR polymerization [Eqs. (2.1) and (2.2)]. These can be modified for the batch case by dropping the flow terms:

$$\frac{dM}{dt} = -k_p MP = -k_p M \sum_{n=1}^{\infty} P_n, \quad M(0) = M_0 \tag{2.5}$$

$$\frac{dP}{dt} = 0, \quad P(0) = P_0 \tag{2.6}$$

In the absence of the flow terms, the reactor volume V appears in all terms and may be canceled. Equation (2.6) simply reflects the fact that, with rapid initiation and no termination, the number of live chains remains constant.

To investigate the molecular weight distribution, it is necessary to develop mass balances over the concentration of live chains of length n (P_n) for each value of n. The balance for P_1 is simply

$$\frac{dP_1}{dt} = -k_p MP_1, \quad P_1(0) = P_{10} \tag{2.7}$$

For P_n, $n > 1$, the balance for each value of n is

$$\frac{dP_n}{dt} = -k_p M(P_n - P_{n-1}), \quad P_n(0) = 0, n \geq 2 \tag{2.8}$$

The initial condition on Eq. (2.8) assumes that no polymer of length greater than 1 is in the reactor at time 0. Equations (2.5)–(2.8) form an infinite set of ordinary differential equations (one for each value of n, plus the monomer balance). An energy balance may be added as well. Numerical solution is simplified in this case because $P(t)$ is a constant, and $M(t)$ can be calculated independent of the P_n equations. The differential equation for each value of n can then be integrated sequentially because $P_n(t)$ depends only on $M(t)$ and $P_{n-1}(t)$. If

$P(t)$ were not constant, the entire system could have been solved (numerically) simultaneously instead of sequentially. However, because common polymerizations reach a degree of polymerization of greater that 1000, the numerical task of such a solution is formidable. In the following sections, three more tractable methods of solution for this and more complex molecular weight problems will be discussed. Differential difference equations such as Eq. (2.8) can solved efficiently in a number of ways. Three techniques will be discussed here: the method of moments, z-transforms, and the continuous variable approximation.

The Method of Moments

Following the nomenclature of Ray [1], if the kth moment of the NCLD is defined as

$$\lambda_k = \sum_{n=1}^{\infty} n^k P_n, \quad k = 0, 1, 2 \tag{2.9}$$

from Eq. (1.1), the NACL may be written as the ratio of the zeroth to the first moments:

$$\mu_n = \frac{\lambda_1}{\lambda_0} \tag{2.10}$$

The variance of the NCLD is the second moment about the mean or

$$\sigma_n^2 = \frac{\lambda_2}{\lambda_0} - \left(\frac{\lambda_1}{\lambda_0}\right)^2 \tag{2.11}$$

Similarly, the kth moment of the WCLD may be written as

$$\hat{\lambda}_k = w \sum_{n=1}^{\infty} n^{(k+1)} P_n, \quad k = 1, 2,\ldots \tag{2.12}$$

Comparison of Eqs. (1.2) and (2.12) will confirm that the WACL may be written as follows:

$$\mu_w = \frac{\hat{\lambda}_1}{\hat{\lambda}_0} = \frac{\lambda_2}{\lambda_1} \tag{2.13}$$

The variance of the WCLD may be written as

$$\sigma_w^2 = \frac{\hat{\lambda}_2}{\hat{\lambda}_0} - \left(\frac{\hat{\lambda}_1}{\hat{\lambda}_0}\right)^2 = \frac{\lambda_3}{\lambda_1} - \left(\frac{\lambda_2}{\lambda_1}\right)^2 \tag{2.14}$$

The means of the NMWD and WMWD may be calculated from the NACL and WACL via Eqs. (1.3) and (1.4). The variances of the NMWD and WMWD are functions of the variances of the NCLD and WCLD, respectively, as follows:

$$\sigma_{mn}^2 = w^2 \sigma_n^2 \tag{2.15}$$

$$\sigma_{mw}^2 = w^2 \sigma_w^2 \tag{2.16}$$

Thus it may be seen that if the leading moments of the NCLD (or WCLD) are known, the mean and the variance of any of the distributions (NCLD, WCLD, NMWD, or WMWD) characterizing the product may be calculated directly. The polydispersity may be calculated from Eq. (1.5) once the NAMW and WAMW have been determined. It is possible to reconstruct the entire distribution from an infinite set of moments [1]. If the distribution is not complex, a good approximation may be made with a finite (even small) number of moments.

In cases of simple kinetics, it may be possible to derive equations for the development of the moments directly from the mass balances. Consider batch anionic solution polymerization with very rapid initiation, and no termination as described by Eqs. (2.5) through (2.8). Equation (2.8) may be multiplied by n^k, added to Eq. (2.7), and then summed for all values of n, resulting in

$$\sum_{n=1}^{\infty} n^k \frac{dP_n}{dt} = -k_p M \sum_{n=1}^{\infty} n^k P_n + k_p M \sum_{n=2}^{\infty} n^k P_{n-1}, \quad \sum_{n=1}^{\infty} P_n = P_{10}$$

$$\tag{2.17}$$

Here

$$\sum_{n=1}^{\infty} n^k P_n = \lambda_k, \quad \sum_{n=1}^{\infty} n^k \frac{dP_n}{dt} = \frac{d\lambda_k}{dt} \tag{2.18}$$

The final term in Eq. (2.17) may be evaluated for $k = 1,2,3$ as

$$k = 0, \quad \sum_{n=2}^{\infty} N^k P_{n-1} = \lambda_0 \tag{2.19}$$

$$k = 1, \quad \sum_{n=2}^{\infty} n^k P_{n-1} = \lambda_0 + \lambda_1 \tag{2.20}$$

$$k = 2, \quad \sum_{n=2}^{\infty} n^k P_{n-1} = \lambda_0 + 2\lambda_1 + \lambda_2 \tag{2.21}$$

Thus the equations for the first three moments become

$$\frac{d\lambda_0}{dt} = 0, \quad \lambda_0(0) = P_{10} \tag{2.22}$$

$$\frac{d\lambda_1}{dt} = k_p M \lambda_0, \quad \lambda_1(0) = P_{10} \tag{2.23}$$

$$\frac{d\lambda_2}{dt} = k_p M \lambda_0 + 2k_p M \lambda_1, \quad \lambda_2(0) = P_{10} \tag{2.24}$$

Integration of Eqs. (2.22) through (2.24), together with the monomer balance [Eq. (2.5)] will yield an adequate characterization of the molecular weight development under isothermal conditions.

z-Transforms

Another way of dealing with the differential difference equations describing the chain length distribution [i.e., Eqs. (3.7) and (3.8)] is through the use of z-transforms. The use of z-transforms is common in digital process control, but can be used to solve systems of difference equations arising from any source. The z-transform of P_n is defined as

$$F(z,t) = \sum_{n=0}^{\infty} z^{-n} P_n(t), \quad P_n(t) = 0 \text{ for } n \leq 0 \tag{2.25}$$

Note that in this case the discrete variable is chain length (n) and that time remains a continuous variable. This is different than in digital control applications where the only independent variable, time, is being discretized. However, standard tables of z-transforms can still be used if this difference is kept in mind and the results are interpreted accordingly. Two properties of z-transforms are important for this application and can be derived from Eqs. (2.9) and (2.25). If the z-transform of P_n is defined by Eq. (2.25), then

$$F(P_{n-k}) = z^{-k} F(P_n) \tag{2.26}$$

The moments of the NCLD can be calculated from the z-transform of $P_n(t)$ as

$$\lambda_k(t) = \lim_{z \to 1} \left\{ (-1)^k \frac{\partial^k F(z,t)}{\partial (\ln z)^k} \right\} \tag{2.27}$$

The z-transform technique is quite similar to the generating function approach [1,2]. It has an advantage, however, in that extensive tables of z-transforms are available in the digital control literature.

For the isothermal batch anionic polymerization represented by Eqs. (2.5) through (2.8), the z-transform technique may be used by first making a variable transformation:

$$\tau = \int_0^t k_p M(t') \, dt' \tag{2.28}$$

This has the effect of removing M from the live chain equations, resulting in

$$\frac{dP_1}{d\tau} = -P_1, \quad P_1(0) = P_{10} \tag{2.29}$$

$$\frac{dP_n}{d\tau} = -(P_n - P_{n-1}), \quad P_n(0) = 0, n \geq 2 \tag{2.30}$$

$$\frac{dM}{d\tau} = -P = -P_{10}, \quad M(0) = M_0 \tag{2.31}$$

Taking the z-transform of Eqs. (2.29) and (2.30), one obtains

$$\frac{dF(z,\tau)}{d\tau} = (z^{-1} - 1)F(z,\tau), \quad F(z,0) = z^{-1} P_{10} \tag{2.32}$$

Equation (2.32) is separable, and can be solved as

$$F(z,\tau) = z^{-1} P_{10} \exp(-\tau) \exp(z^{-1}\tau) \tag{2.33}$$

Expanding in a power series in z^{-1}

$$F(z,t) = P_{10} \exp(-\tau) \sum_{n=1}^{\infty} \frac{(\tau)^{n-1}}{(n-1)!} z^{-n} \tag{2.34}$$

and equating coefficients between Eqs. (2.25) and (2.34) results in

$$P_n(t) = P_{10} \exp(-\tau) \frac{(\tau)^{n-1}}{(n-1)!}, \quad n \geq 1 \tag{2.35}$$

This is a Poisson distribution with mean $(1 + \tau)$ and variance τ.

Continuous Variable Approximation

Equations (2.7) and (2.8) comprise a set of differential difference equations. As noted previously, the set is infinite and must be modified in some way to effect a solution. The difference variable, n, must be a discrete integer because it represents the number of monomer units in a polymer chain. Note, however, that n is a very large number. (A degree of polymerization of 1000 or greater is common.) Because of this, only small errors are introduced by treating n as a continuously variable quantity. If this is done, the infinite set of ordinary differential difference equations can be represented by a single partial differential equation, and known techniques for their solution can be employed.

For the isothermal batch anionic polymerization represented by Eqs. (2.5) through (2.8), this approach may be taken after first making a variable transformation to obtain Eqs. (2.29) and (2.30). If n is then treated as a continuous variable $[P_n(t) = P(n,t)]$, these equations can be written as

$$\frac{\partial P(1,\tau)}{\partial \tau} = -P(1,\tau), \qquad P(1,0) = P_{10} \tag{2.36}$$

$$\frac{\partial P(n,\tau)}{\partial \tau} = -P(n,\tau) + P(n-1,\tau), \qquad P(n,0) = , \quad n \geq 2 \tag{2.37}$$

Expanding $P(n-1,\tau)$ in a Taylor series around $P(n,\tau)$ and truncating after the second-order term gives

$$P(n-1,\tau) = P(n,\tau) - \frac{\partial P(n,\tau)}{\partial n} + \frac{1}{2} \frac{\partial^2 P(n,t)}{\partial n^2}. \tag{2.38}$$

Substituting Eq. (2.38) into Eq. (2.37) gives

$$\frac{\partial P(n,\tau)}{\partial \tau} = -\frac{\partial P(n,\tau)}{\partial n} + \frac{1}{2} \frac{\partial^2 P(n,t)}{\partial n^2} \tag{2.39}$$

with the boundary conditions

$$P(n,0) = P_{10}\delta(n-1) \tag{2.40}$$

$$\lim_{n \to \pm\infty} P(n,\tau) < \infty \tag{2.41}$$

where $\delta(x)$ is the delta function, which has a value of unity when the argument (x) is equal to zero, and a value of zero everywhere else. Equations (2.40) and (2.41) can be solved by standard methods [1] to yield

$$P(n,\tau) = \frac{P_{10}}{(2\pi\tau)^{1/2}} \exp\left\{-\frac{[n-(\tau+1)]^2}{2\tau}\right\} \tag{2.42}$$

This is a Gaussian distribution with mean and variance equal to τ. As n increases, the Gaussian distribution approaches the Poisson distribution, which has been previously shown to be the exact solution. Thus, for large n, the assumption of the continuous variable approximation is valid. Although its use is not as common as that of the z-transform technique, other polymerization systems have been analyzed by the continuous variable approximation, including one report of its incorporation in a closed-loop control scheme [3].

2.1.3 Development and Use of Mathematical Models

As has been demonstrated, the mass and energy balances incorporating the kinetics and reactor type should be combined with a description of the molecular weight development to produce a model of a polymerization reactor which can accurately describe the production rate and character of the product. Often, however, the proper kinetic constants may be unknown. In some cases, even the exact form of the reaction mechanism is not understood. In these cases, it may be necessary to use parameter estimation to fit the model to experimental data. The number of adjustable parameters should be kept low, however, or the model becomes descriptive of particular data sets rather than mechanistic. The reader is referred to a number of excellent works on parameter estimation for reactor models [4–8].

Any model, even if based on established kinetics, should be validated by simulation of data sets which were not used in the estimation of parameters within the model. These data should preferably cover operating regimes far from those of the data used for parameter estimation. Poor agreement suggests incorrect mechanisms or a tendency of the model to simply correlate the data from which its parameters were estimated.

Validated polymerization reactor models may be extremely valuable for design of the reactor system, optimization of the operating parameters, simulation of potential new modes of operation, and even for the development of new products through the modification of the product via modification of the process. In addition, simulation studies can lead to the identification of potential stability problems such as steady-state multiplicity (Chapter 4) and can be used to evaluate potential control schemes.

2.2 Kinetic Analysis

In Chapter 1, a simple kinetic analysis was presented for each polymerization mechanism based on specifying the rate of polymerization and the degree of polymerization. In the present chapter, techniques were discussed for treating the infinite set of species mass balances necessary to describe the entire molecular weight distribution. In this section, the techniques of mathematical modeling will be applied to batch polymerizations of three common types: chain growth without termination (anionic), chain growth with termination (free radical), and step-growth. The emphasis will be on rigorous modeling of the molecular weight distribution (as NCLD) and the effects of operating conditions on the mean (NACL) and breadth of this distribution. In Chapter 3, the analyses will be repeated for polymerization in a CSTR to see the effects of residence time distribution on the NACL.

2.2.1 Anionic Polymerization

Consider batch anionic polymerization as described by Eqs. (1.15)–(1.18). If one assumes very rapid initiation and no termination or chain transfer (living polymer), the kinetics are described by Eqs. (2.5)–(2.8). Direct integration of Eq. (2.5) (P constant) will yield the monomer concentration versus time curve, from which the monomer conversion (x) as a function of time may be calculated:

$$x(t) = \frac{M_0 - M(t)}{M_0} \tag{2.43}$$

Previous analysis of Eqs. (2.7) and (2.8) by z-transform techniques results in Eq. (2.35) which is repeated here for convenience:

$$P_n(t) = P_{10} \exp(-\tau) \frac{(\tau)^{n-1}}{(n-1)!}, \quad n \geq 1 \tag{2.35}$$

where τ is defined by Eq. (2.28):

$$\tau = \int_0^t k_p M(t') \, dt' \tag{2.28}$$

This is a Poisson distribution with mean $(1 + \tau)$ and variance τ.

From Eqs. (1.5), (2.10), and (2.13), it may be seen that the polydispersity, D, may be written as

$$D = \frac{m_w}{m_n} = \frac{\mu_w}{\mu_n} = \frac{\lambda_0 \lambda_2}{\lambda_1^2} \tag{2.44}$$

From Eqs. (2.11) and (2.44), it may be verified that

$$D = 1 + \frac{\sigma_n^2}{\mu_n^2} \tag{2.45}$$

From Eq. (2.28),

$$\mu_n = (1 + \tau) \approx \tau = \sigma_n^2 \tag{2.46}$$

and

$$D = 1 + \frac{\sigma_n^2}{\mu_n^2} = 1 + \frac{1}{\mu_n} \tag{2.47}$$

Thus, for a high degree of polymerization, D approaches unity, meaning the NCLD approaches monodispersity. Hence, ionic polymerization in the absence of termination or chain transfer is useful for creating narrow molecular weight distribution standards. Equation (2.46) can be contrasted with Eq. (1.20) which gives the degree of polymerization. Equation (1.20) presumes that chain transfer is taking place; Eq. (2.46) presumes that it is not. It should be noted that the approach taken in calculating the degree of polymerization (rate of chain growth divided by rate of chain stoppage) is valid only when termination or chain transfer occurs.

2.2.2 Free Radical Polymerization

Consider the free radical polymerization mechanism defined by Eqs. (1.33)–(1.41). Ignoring inhibition [although it may be easily accounted for via Eq. (1.47)] and considering an isothermal batch solution polymerization, the proper mass balances may be written as

$$\frac{dM}{dt} = -k_p PM, \quad M(0) = M_0 \tag{2.48}$$

$$\frac{dI}{dt} = -k_d I, \quad I(0) = I_0 \tag{2.49}$$

Additionally, Eqs. (1.43) and (1.44) may be written to describe the concentrations of primary radicals (R) and total live chains (P). As before, the time derivatives of R and P are set to zero, and R is eliminated from the two equations to give Eq. (1.45):

$$P = \left[\frac{2 f k_d I}{k_{t_c} + k_{t_d}} \right]^{1/2} \tag{1.45}$$

Equations (2.48), (2.49), (1.45), and (1.46) define the conversion-time behavior of the reactor.

Equation (1.48) defines the instantaneous degree of polymerization at any instant during the reaction. However, if complete information about the chain length distributions of the live and dead polymer is desired, balances must be made over each of these species. Following Ray [1] (with modifications), balances over live chains of length 1 and n ($n > 1$), neglecting chain transfer to polymer and the chain transfer agent, result in

$$\frac{dP_1}{dt} = k_i R M_1 - k_p P_1 M_1 + (k_{f_s} S + k_{f_m} M_1)(P - P_1)$$

$$\quad - (k_{t_c} + k_{t_d}) PP_1, \quad P_1(0) = 0 \tag{2.50}$$

$$\frac{dP_n}{dt} = k_p M_1 (P_{n-1} - P_n) - (k_{f_s} S + k_{f_m} M_1) P_n - (k_{t_c} + k_{t_d}) PP_n,$$

$$P_n(0) = 0, \quad n > 1 \tag{2.51}$$

A balance over dead chains of length n $(n > 1)$ can be written as

$$\frac{dM_n}{dt} = (k_{f_s}S + k_{f_m}M_1)P_n + k_{t_d}P_nP + \frac{1}{2} k_{t_c} \sum_{m=1}^{n-1} P_mP_{n-m},$$

$$M_n(0) = 0, \quad n > 1 \tag{2.52}$$

If the QSSA is made for P_1, Eq. (2.50) can be solved to give

$$P_1 = (1 - \alpha)P \tag{2.53}$$

where the probability of propagation is given by

$$\alpha = \frac{k_pM}{k_pM + k_{f_m}M + k_{f_s}S + (k_{t_c} + k_{t_d})P} \tag{2.54}$$

Likewise, making the QSSA for P_n, results in

$$P_n = \alpha P_{n-1} \tag{2.55}$$

P_{n-1} in Eq. (2.55) can be replaced by αP_{n-2}. P_{n-2} can then be replaced with αP_{n-3}. This can be continued until P_1 on the right-hand side of the equation can be replaced from Eq. (2.53), resulting in

$$P_n = (1 - \alpha)P\alpha^{n-1} \tag{2.56}$$

Alternatively, one can take the z-transform of Eqs. (2.53) and (2.55), [that is, multiply Eq. (2.53) by z^{-1}, multiply Eq. (2.55) by z^{-n} and sum over all $n > 1$, then add the two equations], resulting in

$$F_l(z) = z^{-1}(1 - \alpha)P + \alpha z^{-1}F_l(z) \tag{2.57}$$

or

$$F_l(z) = \frac{z^{-1}(1 - \alpha)P}{(1 - \alpha z^{-1})} \tag{2.58}$$

where $F_l(z)$ is the z-transform of the live polymer distribution. Taking the inverse z-transform of Eq. (2.58) gives Eq. (2.56). Note that this is the Flory or "most probable" distribution [10].

Similarly, taking the z-transform of Eq. (2.52) results in

$$\frac{d}{dt}[F_d(z) - z^{-1}M_1] = (k_{f_s}S + k_{f_m}M_1)[F_l(z) - z^{-1}P_1]$$

$$+ k_{t_d}P[F_l(z) - z^{-1}P_1] + \frac{1}{2} k_{t_c}F_l^2(z) \tag{2.59}$$

where $F_d(z)$ is the z-transform of the dead polymer distribution.

The moments of the live and dead NCLD can now be evaluated. Equation (2.27) may be applied to Eq. (2.58) to obtain the moments of the live polymer NCLD:

$$\lambda_0 = P \tag{2.60}$$

$$\lambda_1 = \frac{P}{1 - \alpha} \tag{2.61}$$

$$\lambda_2 = \frac{P(1 + \alpha)}{1 - \alpha} \tag{2.62}$$

Using Eqs. (2.10), (2.13), and (1.5), the NACL, WACL, and polydispersity can be calculated from the moments:

$$\mu_n = \frac{\lambda_1}{\lambda_0} = \frac{1}{1 - \alpha} \qquad \text{(live polymer)} \tag{2.63}$$

$$\mu_w = \frac{\lambda_2}{\lambda_1} = \frac{1 + \alpha}{1 - \alpha} \qquad \text{(live polymer)} \tag{2.64}$$

$$D = \frac{m_w}{m_n} = \frac{\mu_w}{\mu_n} = 1 + \alpha \qquad \text{(live polymer)} \tag{2.65}$$

For long chains, the probability of propagation approaches unity so that

$$D = (1 + \alpha) \approx 2 \qquad \text{(live polymer)} \tag{2.66}$$

The moments of the dead polymer NCLD may now be calculated. Because we wish to exclude monomer (M_1) from the distribution, the moments of the dead polymer NCLD are defined as

$$\eta_k = \sum_{n=2}^{\infty} n^k M_n, \qquad k = 0, 1, 2,... \tag{2.67}$$

For the moments defined above, the analog to Eq. (2.27) is

$$\eta_k(t) = \lim_{z \to 1} \left\{ (-1)^k \frac{\partial^k F^*(z, t)}{\partial (\ln z)^k} \right\} \tag{2.68}$$

where

$$F^*(z) = [F_d(z) - z^{-1} M_1] \tag{2.69}$$

When Eqs. (2.68) and (2.69) are used to differentiate Eq. (2.59), the time derivatives of the leading moments may be written as

$$\frac{d\eta_0}{dt} = (k_{fm}M + k_{td}P + k_{fs}S)P\alpha + 0.5k_{tc}P^2, \qquad \eta_0(0) = 0 \qquad (2.70)$$

$$\frac{d\eta_1}{dt} = \frac{[(k_{fm}M + k_{td}P + k_{fs}S)(2\alpha - \alpha^2) + k_{tc}P]P}{1 - \alpha}, \qquad \eta_1(0) = 0$$

$$(2.71)$$

$$\frac{d\eta_2}{dt} = \frac{[(k_{fm}M + k_{td}P + k_{fs}S)(\alpha^3 - 3\alpha^2 + 4\alpha) + k_{tc}P(\alpha + 2)]P}{(1 - \alpha)^2},$$

$$\eta_2(0) = 0 \qquad (2.72)$$

To derive the NCLD for the dead polymer, Eqs. (2.70)–(2.72) must be integrated numerically. If, however the temperature, monomer concentration and initiator concentration are all assumed to be constant, a lower bound for the dead polymer polydispersity may be calculated. In this case, the right-hand sides of Eqs. (2.70)–(2.72) are constant, and the integration of each is trivial. The NACL and WACL of the dead polymer distribution can then be calculated as follows:

$$\mu_n = \frac{\eta_1}{\eta_0} = \frac{(k_{fm}M_1 + k_{td}P + k_{fs}S)(2\alpha - \alpha^2) + k_{tc}P}{(1 - \alpha)[(k_{fm}M_1 + k_{td}P + k_{fs}S)\alpha + 0.5k_{tc}P]} \qquad (2.73)$$

$$\mu_w = \frac{\eta_2}{\eta_1} = \frac{(k_{fm}M_1 + k_{td}P + k_{fs}S)(\alpha^3 - 3\alpha^2 + 4\alpha) + k_{tc}P(\alpha + 2)}{(1 - \alpha)[(k_{fm}M_1 + k_{td}P + k_{fs}S)(2\alpha - \alpha^2) + k_{tc}P]}$$

$$(2.74)$$

Substituting Eqs. (2.53) and (2.56) into Eq. (2.52) results in

$$\frac{dM_n}{dt} = [k_{fs}S + k_{fm}M_1 + k_{td}P](1 - \alpha)P\alpha^{n-1}$$

$$+ 0.5k_{tc}P^2(1 - \alpha)^2\alpha^{n-2}(n - 1), \quad M_n(0) = 0, \quad n > 1 \quad (2.75)$$

Assuming the right-hand side of Eq. (2.75) constant and integrating yields

$$M_n(t) = t \left\{ \begin{array}{l} [k_{fs}S + k_{fm}M_1 + k_{td}P](1 - \alpha)P\alpha^{n-1} \\ + 0.5k_{tc}P^2(1 - \alpha)^2\alpha^{n-2}(n - 1), \end{array} \right\} \quad M_n(0) = 0,$$

$$n > 1 \qquad (2.76)$$

Two special cases may now be considered.

If the long chain approximation is made, and there is no termination by combination,

$$M_n(t) = \eta_0(1 - \alpha)\alpha^{n-2} \tag{2.77}$$

The power of $n - 2$ comes about because in this treatment we have excluded M_1 from consideration as a polymer. If M_1 (here meant to indicate polymer of length unity formed by a termination or transfer reaction from P_1, and not monomer) is included in the normalization, the distribution may be written as

$$M_n(t) = M_{tot}(1 - \alpha)\alpha^{n-1} \tag{2.78}$$

where M_{tot} is taken to mean all polymer of length 1 or greater. This is again the Flory distribution and is exactly the distribution defined by Eq. (2.56). Thus, for termination by disproportionation only, the dead polymer will have the same distribution as the live polymer. Assuming the probability of propagation to be unity, the polydispersity of the dead polymer may be calculated as the ratio Eq. (2.74) divided by Eq. (2.73):

$$D = \frac{\mu_w}{\mu_n} \approx 2 \quad \text{(dead polymer)} \tag{2.79}$$

From Eqs. (2.63) and (2.73), it may be seen that

$$\frac{\mu_n(\text{dead})}{\mu_n(\text{live})} = 2 - \alpha \approx 1 \tag{2.80}$$

If, instead, termination by combination is assumed to be the only mode of chain termination (no chain transfer to solvent or monomer, and no termination by disproportionation),

$$M_n(t) = \eta_0(n - 1)(1 - \alpha)^2 \alpha^{n-2} \quad \text{(dead polymer)} \tag{2.81}$$

Because no M_1 chains can be formed by combination, M_{tot} is equal to η_0, and Eq. (2.81) is equivalent to both Eqs. (2.77) and (2.78). The polydispersity, for long chains, is

$$D = \frac{\mu_w}{\mu_n} = \frac{2 + \alpha}{2} \approx 15 \quad \text{(dead polymer)} \tag{2.82}$$

From Eqs. (2.66) and (2.73), it may be seen that

$$\frac{\mu_n(\text{dead})}{\mu_n(\text{live})} = 2 \tag{2.83}$$

and the dead chains have twice the NACL of the live chains.

Thus, under ideal conditions, the polydispersity will range from 1.5 for termination by combination and no chain transfer, to 2.0 for termination by disproportionation. Polydispersities above 2.0 occur due to variations in operating conditions (temperature, monomer or initiator concentration, etc.) with time or other phenomena such as the gel effect or poor mixing. Because the monomer and initiator concentrations are not constant in a batch reactor, this analysis represents only a limiting case.

An Example

As an example of mathematical modeling, let us consider the semibatch solution free radical polymerization of methyl methacrylate [9]. Balances over monomer, initiator, and solvent yield, respectively,

$$\frac{d(VM)}{dt} = -Vk_pPM, \qquad M(0) = M_0 \tag{2.84}$$

$$\frac{d(VI)}{dt} = QI_f - Vk_dI, \qquad I(0) = I_0 \tag{2.85}$$

$$\frac{d(VS)}{dt} = QS_f, \qquad S(0) = S_0 \tag{2.86}$$

Note that this model allows initiator dissolved in solvent to be added to the reactor over time. Thus, inflow terms appear in these balances. Note also that because of the semibatch nature, the reactor volume must remain inside the differentiation. To account for the change in volume, a balance is written on the reactor volume:

$$\frac{dV}{dt} = Q + \epsilon \frac{dM}{dt}, \qquad V(0) = V_0 \tag{2.87}$$

Here q is the volumetric flow rate of the initiator stream, and ϵ is the change in reactor volume per mole of monomer polymerized. An energy balance can be included as

$$\frac{d(V\rho C_p T)}{dt} = V(-\Delta H_p)k_pPM - UA(T - T_j), \qquad T(0) = T_0 \tag{2.88}$$

The first three moments of the dead polymer NCLD can be found in a manner analogous to Eqs. (2.70)–(2.72). The difference is that in the semibatch case, reaction volume V, must remain inside the differentiation:

$$\frac{d(V\eta_0)}{dt} = (k_{fm}M + k_{td}P + k_{fs}S)PV\alpha + 0.5k_{tc}P^2V, \qquad \eta_0(0) = 0 \tag{2.89}$$

$$\frac{d(V\eta_1)}{dt} = \frac{[(k_{fm}M + k_{td}P + k_{fs}S)(2\alpha - \alpha^2) + k_{tc}P]VP}{(1 - \alpha)},$$

$$\eta_1(0) = 0 \tag{2.90}$$

$$\frac{d(V\eta_2)}{dt} =$$

$$= \frac{[(k_{f_m}M + k_{f_d}P + k_{f_s}S)(\alpha^3 - 3\alpha^2 + 4\alpha) + k_{t_c}P(\alpha + 2)]VP}{(1-\alpha)^2},$$

$$\eta_2(0) = 0 \tag{2.91}$$

To complete the model, the gel and glass effect correlations of Schmidt and Ray [11] are used to predict the effect of viscosity on the propagation and termination rate constants, and P is calculated from Eq. (1.45). The model, defined by Eqs. (2.84)–(2.91) and (1.45), may now be nondimensionalized [9] and integrated using a standard numerical integration routine. The nondimensionalization reduces the computational time for the integration by as much as an order of magnitude. The results of such a simulation for the batch (no initiator feed during reaction) polymerization of methyl methacrylate initiated with benzoyl peroxide in 40% ethyl acetate at 70 °C are given in Fig. 2.1. Note that the monomer conversion rises steadily to 85% where the reaction has been stopped arbitrarily while the monomer concentration falls correspondingly. Initiator concentration is almost constant because initiator decomposition is quite slow. This is often the case and can sometimes lead to simplifications of the kinetic analysis. The reactor volume goes down due the reduction in volume on polymerization, and the solvent concentration goes up slightly due to the volume reduction on polymerization. Note that the NAMW remains fairly constant until the gel effect becomes significant, at which time it begins to rise because termination is more strongly retarded by viscosity than propagation. Eventually, the propagation reaction becomes retarded, and the NAMW begins to fall again near the end of the reaction.

2.2.3 Step-Growth Polymerization

A–B Type Step-Growth Polymerization

Recall from Chapter 1 that if only one monomer is involved in a step-growth polymerization, the monomer must contain both functional groups necessary for the condensation reaction. If the monomer molecule is represented as $A–B$, the condensation of two monomer molecules may be written as

$$A-B + A-B \overset{k_p}{\underset{k_p'}{\rightleftharpoons}} A-B-A-B + W \tag{2.92}$$

where W is the condensation by-product (usually water). The dimer will now react with any other monomer or oligomer unit, and the resulting chain will condense with another chain to form longer and longer polymer chains. This repeated condensation may be represented by

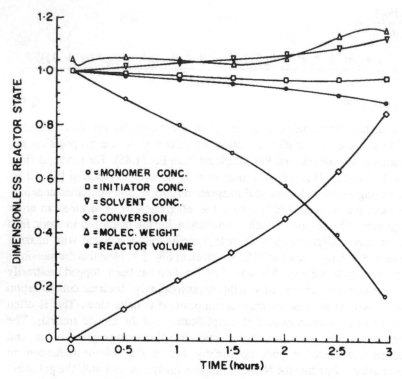

Figure 2.1 Isothermal batch solution polymerization. Monomer, initiator, and solvent concentrations, and reactor volume are normalized by their initial values: molecular weight is normalized by division by a target molecular weight of 150,000. (From Ref. 9.)

$$P_n + P_m \overset{k_p}{\underset{k_p'}{\rightleftharpoons}} P_{n+m} + W \tag{2.93}$$

If the polymerization is carried out in a batch reactor with provisions for by-product removal (usually by boiling off the by-product water), the mechanism can be considered irreversible, and the following kinetic equations can be written:

$$\frac{dP_1}{dt} = -k_p P_1 P, \qquad P_1(0) = P_{10} \tag{2.94}$$

$$\frac{dP_n}{dt} = \frac{1}{2} k_p \sum_{r=1}^{n-1} P_r P_{n-r} - k_p P_n P, \qquad P_n(0) = P_{n0} \qquad n > 1 \tag{2.95}$$

$$P = \sum_{n=1}^{\infty} P_n \tag{2.96}$$

Multiplying Eq. (2.94) by z^{-1}, and Eq. (2.95) by z^{-n}, and summing over all values of n results in

$$\frac{dF(z,t)}{dt} = \frac{1}{2}k_pF^2(z,t) - k_pP(t)F(z,t), \quad F(z,0) = f_0(z)$$

$$= \sum_{n=1}^{\infty} z^{-n}P_{n0} \tag{2.97}$$

where $P(t) = F(1,t)$ is given by

$$\frac{dP(t)}{dt} = -\frac{1}{2}k_pP^2(t), \quad P(0) = f_0(1) = \sum_{n=1}^{\infty} P_{n0} \tag{2.98}$$

Transforming the time variable as was done earlier,

$$\tau = \frac{1}{2}\int_0^t k_p(t')\,dt' \tag{2.99}$$

Equations (2.97) and (2.98) become

$$\frac{dF(z,\tau)}{d\tau} = F^2(z,\tau) - 2P(\tau)F(z,\tau), \quad F(z,0) = f_0(z) \tag{2.100}$$

$$\frac{dP(\tau)}{d\tau} = -P^2(\tau), \quad P(0) = f_0(1) \tag{2.101}$$

Equation (2.101) may be solved as

$$P(\tau) = \frac{f_0(1)}{f_0(1)\tau + 1} \tag{2.102}$$

Substituting $y = F(z,\tau)/P(\tau)$, Eq. (2.100) may be written as

$$\frac{dy}{dP} = \frac{1}{P}y(1-y), \quad y[f_0(1)] = \frac{f_0(z)}{f_0(1)} \tag{2.103}$$

which may be solved by separation of variables to give

$$F(z,\tau) = Py = \frac{f_0(z)[P/f_0(1)]^2}{1 - [1 - P/f_0(1)][f_0(z)/f_0(1)]} \tag{2.104}$$

Assuming pure monomer charged to the reactor, $f_0(z) = z^{-1}P_{10}$ and Eq. (2.104) can be expanded as

$$F(z,\tau) = P_{10} \sum_{n=1}^{\infty} \left(\frac{P}{P_{10}}\right)^2 \left(1 - \frac{P}{P_{10}}\right)^{n-1} z^{-n} \tag{2.105}$$

Comparing this with the definition of the z-transform yields the NCLD

$$P_n(\tau) = P_{10}\left(\frac{P}{P_{10}}\right)^2 \left(1 - \frac{P}{P_{10}}\right)^{n-1} = \frac{1}{P_{10}\tau^2}\alpha^{n+1}, \quad n \geq 1 \tag{2.106}$$

$$\alpha = \frac{P_{10}\tau}{P_{10}\tau + 1} \tag{2.107}$$

This is the most probable distribution of Flory.

If Eq. (2.102) is used to substitute for P, Eq. (2.106) becomes

$$P_n(\tau) = \frac{P_{10}(P_{10}\tau)^{n-1}}{(P_{10}\tau + 1)^{n+1}}, \quad n \geq 1 \tag{2.108}$$

The moments of the distribution (excluding P_1) can be calculated from Eqs. (2.68) and (2.104), and the NACL, WACL, and polydispersity can be calculated from the moments as before:

$$\mu_n = \frac{\lambda_1}{\lambda_0} = \frac{2 - \alpha}{1 - \alpha} \tag{2.109}$$

$$\mu_w = \frac{\lambda_2}{\lambda_1} = 1 + \frac{2}{(1 - \alpha)(2 - \alpha)} \tag{2.110}$$

$$D = \frac{\mu_w}{\mu_n} = \frac{1 - \alpha}{2 - \alpha} + \frac{2}{(2 - \alpha)^2} \tag{2.111}$$

Under conditions of complete conversion of monomer to polymer ($\alpha \approx 1$), the polydispersity reduces to a minimum value of 2.

The mole fraction of polymer of chain length n is

$$\frac{P_n(\tau)}{P(\tau)} = \frac{\alpha^n}{P_{10}\tau} \tag{2.112}$$

Recall that the extent of reaction is defined as the fraction of A or B functional groups which have reacted:

$$p = \frac{P_{10} - P}{P_{10}} \tag{2.113}$$

Substituting Eq. (2.102) for P (assuming a pure monomer initial charge) yields

$$p = \frac{P_{10}\tau}{P_{10}\tau + 1} = \alpha \tag{2.114}$$

Thus, both the NCLD [from Eq. (2.106)] and the extent of reaction [from Eq. (2.114)] are fixed for a fixed value of $P_{10}\tau$. Conversely, there is a unique NCLD for each extent of reaction in a batch reactor, even under nonisothermal conditions. For values of α close to 1 (necessary to produce high NACL), Eq. (2.109) is approximately equal to Eq. (1.53) ($\alpha = p$). The difference is due to the exclusion of monomer from the NACL in Eq. (2.109). This becomes insignificant at the high extent of reaction.

A-A/B-B Type Step-Growth Polymerization

Recall from Chapter 1 that if two monomers are involved in a step-growth polymerization, and if each monomer contains two identical functional groups, the polymerization is denoted as $A-A/B-B$. The condensation reaction then takes place between the A and B functional groups on adjacent monomer units. If the monomer molecules are represented by $A-A$ and $B-B$, three polymeric species will be present at any time. If these are represented as

$(A-A-B-B)_{n-1}A-A,\quad A_n$

$(A-A-B-B)_n,\quad M_n$

$(B-B-A-A)_{n-1}B-B,\quad B_n$

the three possible polymerization reactions may be written as

$$A_n + M_m \underset{k_p'}{\overset{k_p}{\rightleftharpoons}} A_{n+m} + W \tag{2.115}$$

$$A_n + B_m \underset{k_p'}{\overset{k_p}{\rightleftharpoons}} M_{n+m-1} + W \tag{2.116}$$

$$B_n + M_m \underset{k_p'}{\overset{k_p}{\rightleftharpoons}} B_{n+m} + W \tag{2.117}$$

where W is the condensation by-product. An analysis similar to that of Section 4.3.1 may be carried out based on these reactions [2] but is beyond the scope of this book. If the stoichiometric ratio is unity, the results will be similar to those above. If the stoichiometric ratio deviates significantly ratio is unit, the maximum molecular weight is reduced as shown in Fig. 1.8.

2.3 Copolymerization

The simultaneous polymerization of two or more monomers results in a macromolecule in which the two or more types of monomer unit are distributed in some fashion throughout the polymer structure. The degree of distribution of the monomer units may range from a strictly alternating structure to a graft copolymer in which chains of polymer B are grafted onto chains of polymer A. The various monomer units must, however, appear within the same polymer molecule. Copolymers are thus distinguished from physical blends of polymers. The ability to create a macromolecule containing two or more types of monomer units gives the polymer chemist a greatly increased ability to custom design a polymer to yield specific end-use properties.

2.3.1 Copolymer Structure

The various possible copolymer structures are shown in Table 2.1. If units of monomers A and B occur in a strictly alternating sequence along the backbone of the polymer chain, the copolymer is said to be *alternating*. If monomers A and B appear randomly along the polymer chain, the copolymer is said to be *random*. As a general rule, both of these copolymers will have properties intermediate between those of the two homopolymers, with the overall copolymer composition (percent A units in the polymer) determining whether the copolymer more nearly resembles homopolymer A or B. Small amounts (1–5% by weight) of comonomer B may, however, have a great effect on the properties of the copolymer when compared with a homopolymer of A. If the A and B units in the copolymer exist in sequences of pure A followed by sequences of B, the material is termed a *block* copolymer. If chains of homopolymer B are attached as appendages to a backbone of homopolymer A, the resulting structure is termed a *graft* copolymer. Both block and graft copolymers are likely to have properties different from those of either homopolymer because there are, in essence, domains of A polymer interspersed with domains of B polymer. This structure may be exploited to develop polymers possessing unique properties not obtainable with homopolymers or with alternating or random copolymers.

A copolymer can be described by specifying three sets of statistics: a molecular weight distribution, the copolymer composition distribution, and the copolymer sequence distributions. The molecular weight can be described by any of the chain length distributions described previously (NCLD, etc.) and can be calculated in much the same ways as for homopolymers, although the calculations are more complex. The copolymer composition distribution is the distribution of chains (x_y) containing various fractions (y) of monomer A (or B). The mean of this distribution is the average copolymer composition. For a binary copolymer, this is the fraction of monomer A (or B) in the polymer which would be found by analysis of a bulk sample of the polymer.

Table 2.1 Copolymer structure.

Alternating	$-A-B-A-B-A-B-A-B-A-B-A-B-$
Random	$-B-A-A-B-A-B-B-B-A-A-B-A-$
Block	$-B-A-A\bullet\bullet\bullet A-A-B-B\bullet\bullet\bullet B-B-A-A-$
Graft	$-A-A-A\bullet\bullet\bullet A-A-A\bullet\bullet\bullet A-A-A\bullet\bullet\bullet A-A-A-$

```
Graft      -A-A-A•••A-A-A•••A-A-A•••A-A-A-
               |       |       |       |
               B       B       B       B
               B       B       B       B
               B       B       B       B
               |       |       |       |
```

Because the sequence of monomer units in a chain strongly affects the properties of the polymer, it is important to describe the distribution of monomer sequences. This will allow one to distinguish between alternating, random, and block copolymers. The copolymer sequence distribution, then, is the distribution of sequences of n units of monomer A or B (A_n, B_n) plotted versus n. Thus, for a strictly alternating binary copolymer, the copolymer composition distributions for both monomers will be delta functions (spikes) at $n = 1$. The copolymer composition distribution and copolymer sequence distributions are shown schematically in Fig. 2.2. In a random copolymer, off-spec copolymer composition distribution will result in poor mechanical properties due to inhomogeneity or improper monomer ratio in the polymer. Off-spec copolymer sequence distribution in a block copolymer may result in poor mechanical properties due to improper lengths and/or distribution of hard (stiff, high glass-transition temperature) and soft (flexible, low glass-transition temperature) segments.

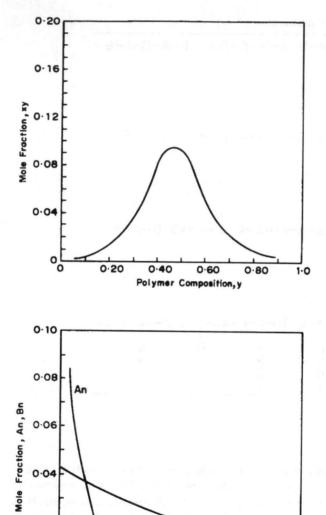

Figure 2.2 Copolymer composition (top) and sequence distributions (bottom).

2.3.2 Chain-Growth Copolymerization

The kinetics of chain-growth copolymerization are best illustrated through the derivation of the copolymerization equation. Consider free radical polymerization of two monomers A and B [12]. The four possible propagation reactions can be written as follows:

$$\left. \begin{array}{l} P_{n,m} + A \overset{k_{paa}}{\Longrightarrow} P_{n+1,m} \\[2mm] P_{n,m} + B \overset{k_{pab}}{\Longrightarrow} Q_{n,m+1} \\[2mm] Q_{n,m} + A \overset{k_{pba}}{\Longrightarrow} P_{n+1,m} \\[2mm] Q_{n,m} + B \overset{k_{pbb}}{\Longrightarrow} Q_{n,m+1} \end{array} \right\} \tag{2.118}$$

Here $P_{n,m}$ is a live chain containing n monomer A units and m monomer B units and having an A unit at the radical end. Likewise, $Q_{n,m}$ is a live chain containing n monomer A units and m monomer B units and having a B unit at the radical end. If the long chain hypothesis is made (propagation occurs much more frequently than chain initiation or termination) and balances over A and B are written, the equation expressing the instantaneous rate of consumption of A relative to B can be written as

$$\frac{dA}{dB} = \frac{k_{paa}PA + k_{pba}QA}{k_{pab}PB + k_{pbb}QB} \tag{2.119}$$

Here P is taken to mean the total concentration of live chains ending in an A unit. Q is the total concentration of live chains ending in a B unit. If the QSSA is made for P and Q, it follows that the rate of conversion of P to Q equals the rate of conversion of Q to P. Using this assumption, and defining the *reactivity ratios*

$$r_1 = \frac{k_{paa}}{k_{pab}}, \quad r_2 = \frac{k_{pbb}}{k_{pba}} \tag{2.120}$$

the *copolymerization equation* results:

$$\frac{dA}{dB} = \frac{A(r_1A + B)}{B(A + r_2B)} \tag{2.121}$$

This may be written as

$$F_1 = \frac{r_1f_1^2 + f_1f_2}{r_1f_1^2 + 2f_1f_2 + r_2f_2^2} \tag{2.122}$$

where F_1 is the mole fraction A in the copolymer being formed at the current instant (instantaneous copolymer composition) and f_1 and f_2 are the instantaneous mole fractions of A and B, respectively, in the monomer mix. F_1 represents the *average* instantaneous copolymer composition. This is adequate because the distribution of instantaneous copolymer compositions has been shown to be quite narrow [13].

The copolymerization equation can be used to understand the various copolymer structures. If $r_1 = r_2 = 0$, the structure is strictly alternating and

$$F_1 = \frac{f_1 f_2}{2 f_1 f_2} = \frac{1}{2}, \quad \frac{dA}{dB} = 1 \tag{2.123}$$

If $r_1 = 1/r_2$ ($r_1 r_2 = 1$), the structure is random because P and Q chains have an equal likelihood of adding the other monomer.

$$\frac{dA}{dB} = r_1 \frac{A}{B} \tag{2.124}$$

If r_1 and r_2 are greater than unity, a block copolymer results, because there is little likelihood of cross-polymerization. For most copolymerization systems, $0 < r_1 r_2 < 1$ and the copolymer is somewhere between alternating and random.

Figures 2.3 and 2.4 show F_1 as a function of f_1 for a number of ratios of r_1/r_2. It should be emphasized that F_1 represents the *instantaneous* copolymer composition. As a batch polymerization progresses, f_1 will change as the more reactive monomer is used up more rapidly, and so chains formed early in the polymerization will contain larger fractions of the more reactive monomer, whereas those formed later will have a higher fraction of the less reactive monomer. Compensation for this effect can be accomplished by adding the more reactive monomer over time in a semibatch mode (programmed comonomer addition). Figure 2.5 indicates that the copolymer composition can be very strongly influenced by the polymerization system employed. Shown in the figure is the instantaneous copolymer composition versus monomer composition in the feed for the CSTR copolymerization of styrene and methyl acrylate. From Fig. 2.5, it may be seen that for homogeneous polymerization (bulk or solution), the polymer will have approximately the same composition as the feed. However, when the polymerization is carried out in a heterogeneous (emulsion) polymerization system, the styrene is polymerized at a much greater rate than the methyl acrylate. This is due to the fact that the polymerization occurs in the styrene-rich polymer particles whereas the more water-soluble methyl acrylate remains dissolved in the aqueous phase. (See Chapter 4.)

The copolymer sequence distribution for a free radical system can be easily calculated. Following Ray [1] and referring to the polymerization system in Eq. (2.118), the copolymer sequence distribution may be determined as follows: Define

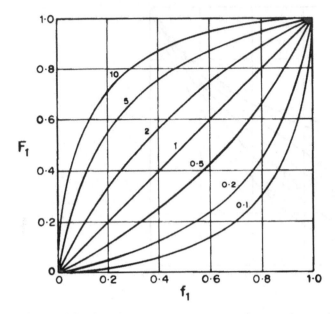

Figure 2.3 Copolymer composition — random copolymers. Instantaneous copolymer composition (F_1) as a function of monomer composition (f_1) for random copolymerization with the values of $r_1 = 1/r_2$ indicated. (From Ref. 14 with permission.)

L_{ij} as the probability that a growing chain ending in an i monomer unit ($i = A$ or B) will add a j ($j = A$ or B) monomer unit next. From this definition, it may be seen that

$$L_{aa} + L_{ab} = L_{ba} + L_{bb} = 1 \tag{2.125}$$

The probability of having exactly n units of A in a series in a growing chain is then

$$A_n = L_{aa}^{n-1}L_{ab} \tag{2.126}$$

Because

$$\sum_{n=1}^{\infty} A_n = L_{ab} \sum_{n=0}^{\infty} L_{aa}^n = \frac{L_{ab}}{1 - L_{aa}} = 1 \tag{2.127}$$

A_n is also the fraction of all A sequences having a length of n. For a batch copolymerization following Eq. (5.1),

$$L_{aa} = \frac{k_{paa}A}{k_{paa}A + k_{pab}B} = \frac{r_1(A/B)}{r_1(A/B) + 1} \tag{2.128}$$

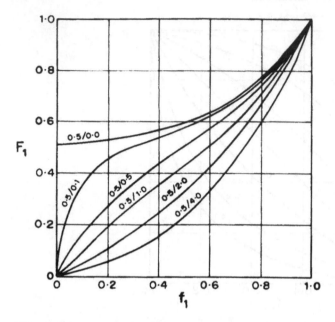

Figure 2.4 Copolymer composition. Instantaneous copolymer composition (F_1) as a function of monomer composition (f_1) for copolymerization with the values of reactivity ratio r_1/r_2 indicated. (From Ref. 14 with permission.)

$$L_{ab} = \frac{k_{pab}B}{k_{paa}A + k_{pab}B} = \frac{1}{r_1(A/B) + 1} \qquad (2.129)$$

$$L_{ba} = \frac{k_{pba}A}{k_{pbb}B + k_{pba}A} = \frac{1}{r_2(B/A) + 1} \qquad (2.130)$$

$$L_{bb} = \frac{k_{pbb}B}{k_{pbb}B + k_{pba}A} = \frac{r_2(B/A)}{r_2(B/A) + 1} \qquad (2.131)$$

Putting Eqs. (2.128) and (2.129) into Eq. (2.126) and expressing the result in terms of monomer mole fraction f_1 gives

$$A_n = \frac{\alpha^{n-1}}{(\alpha + 1)^n} \qquad (2.132)$$

where

$$\alpha = \frac{r_1 f_1}{1 - f_1} \qquad (2.133)$$

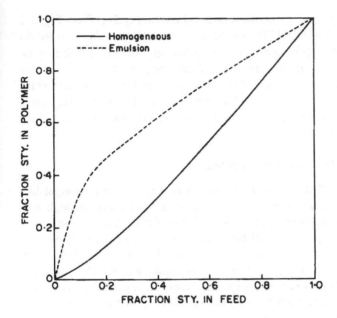

Figure 2.5 Copolymer composition — heterogeneous polymerization. Instantaneous copolymer composition as a function of monomer composition in the feed, for the CSTR copolymerization of styrene and methyl acrylate. (From Ref. 15 with permission.)

Similarly,

$$B_n = \frac{\beta^{n-1}}{(\beta + 1)^n} \qquad (2.134)$$

where

$$\beta = \frac{r_2(1 - f_1)}{f_1} \qquad (2.135)$$

The average sequence lengths can be determined from the distributions:

$$\bar{N}_A = \sum_{n=1}^{\infty} nA_n = \frac{1}{\alpha + 1} \sum_{n=0}^{\infty} (n + 1)\left(\frac{\alpha}{\alpha + 1}\right)^n = 1 + \alpha \qquad (2.136)$$

$$\bar{N}_B = \sum_{n=1}^{\infty} nB_n = -\frac{1}{\beta + 1} \sum_{n=0}^{\infty} (n + 1)\left(\frac{\beta}{\beta + 1}\right)^n = 1 + \beta \qquad (2.137)$$

Representative values of these distributions are plotted in Fig.2.2. From Fig. 2.2 and from Eqs. (2.136) and (2.137), it is apparent that a large value of α (or β) will result in a broad distribution of sequence lengths, some of which may be quite long (a high degree of "blockiness"). Conversely, a small value of α (or β) will result in a very narrow distribution of sequence lengths. A high value of α and a low value of β will result in long sequences of A with short sequences of B interspersed. In the limit ($\alpha \implies 0$ and $\beta \implies 0$), the result will be a perfect alternating polymer ($r_1 = r_2 = 0$).

2.3.3 Step-Growth Copolymerization

The reader will recall that to develop a reasonably high molecular weight in a step-growth polymerization, very high conversions are necessary. At high conversion, the average copolymer composition will approximate the initial monomer composition. The copolymer composition distribution will be reasonably narrow, with a mean at or near the composition of the original monomer mixture. Thus, copolymer composition distribution is not a significant variable affecting copolymer quality. The sequence distribution, however, is of major importance because the desirable properties of many step-growth copolymers depend on the degree of blockiness.

For an A–B type copolymerization (the copolymerization of monomer A–B with another monomer, A'–B'), the sequence distribution may be analyzed using the approach outlined above for free radical copolymerization. The reactivity ratios tend, however, to approach unity, giving an essentially random copolymer.

An A–A/B–B copolymerization (the copolymerization of monomers A–A, B–B, A'–A', and B'–B') offers an additional degree of freedom. An alternating polymer ($-A$–A–B–B–A–A–B'–B'–$) may be made by prepolymerizing to form A–A–B–B–A–A in a separate reaction, then polymerizing the prepolymer in the presence of the B'–B' monomer to form a strictly alternating structure. If a random copolymer is desired, the four monomers are mixed, then polymerized. The sequence distribution is then governed by probability considerations as in free radical copolymerization. If a block copolymer is desired, the blocks are prepared as prepolymers, then joined in the copolymerization reactor.

2.3.4 Graft Copolymers

Graft copolymers are formed by grafting of a secondary polymer (or copolymer) onto a primary, or backbone polymer (or copolymer). In free radical polymerization, this often involves chain transfer to polymer (Chapter 1). Free radical graft copolymerization should not be confused with the polymerization of a branched polymer (or copolymer) structure formed by chain transfer to polymer during initial polymerization. Grafting involves the formation of a polymer in which the branches have a significantly different chemical structure from that of the backbone polymer and is carried out as a post-polymerization process.

Grafting of step-growth polymers is done with the use of trifunctional monomers in the backbone polymer. Two of the functional groups are used to form the backbone polymer. A second homopolymer (or copolymer) is formed in a separate reaction, and the two are joined through the unreacted functional group on the backbone. To avoid branching in the backbone polymer, the third functional group must be of significantly lower reactivity than the other two. Alternatively, a macromer is used. A macromer is an oligomer with a double-bond end group. The backbone polymer copolymerizes with the macromer, forming a graft copolymer with very uniform (in chemical structure and molecular weight) branches.

References

1. W. H. Ray, "On the Mathematical Modeling of Polymerization Reactors," *J. Macromol. Sci.—Revs. Macromol. Chem.*, *C8*, 1–56 (1972).
2. M. V. Tirrell and R. L. Laurence, "Polymerization Reaction Engineering," notes for a short course at the University of Minnesota Polymerization and Polymer Process Engineering Center, Minneapolis, MN, 1983.
3. V. Gonzalez, T. W. Taylor, and K. F. Jensen, "On-Line Estimation of Molecular Weight Distributions in Methyl Methacrylate Polymerization," *Proceedings of the 1986 American Control Conference* (1986).
4. G. E. P. Box and W. G. Hunter, "A Useful Method for Model-Building," *Technometrics*, *4*, 301 (1962)
5. J. R. Kittrell, W. G. Hunter, and C. C. Watson, "Obtaining Precise Estimates for Nonlinear Catalytic Rate Models," *AIChE J.*, *12*, 5 (1966).
6. A. C. Atkinson and W. G. Hunter, "The Design of Experiments for Parameter Estimation," *Technometrics*, *10*, 271 (1968).
7. W. G. Hunter, W. J. Hill, and T. L. Henson, "Designing Experiments for Precise Estimation of All or Some of the Constants in a Mechanistic Model," *Can. J. Chem. Eng.*, *47*, 76 (1969).
8. C. Kiparissides, J. F. MacGregor, and A. E. Hamielec, "Continuous Emulsion Polymerization of Vinyl Acetate, Part II, Parameter Estimation and Simulation Studies," *Can. J. Chem. Eng.*, *58*, 56 (1980).
9. W. E. Houston and F. J. Schork, "Adaptive Predictive Control of a Semibatch Polymerization Reactor," *Polym. Process Eng.*, *5*, No. 1, 119–147 (1987).
10. P. Flory, *J. Amer. Chem. Soc.*, *58*, 1877 (1936).
11. A. D. Schmidt and W. H. Ray, "The Dynamic Behavior of Continuous Polymerization Reactors—I, Isothermal Solution Polymerization in a CSTR," *Chem. Eng. Sci.*, *36*, 1401 (1981).
12. G. E. Ham, "General Aspects of Free-Radical Polymerization," in *Kinetics and Mechanisms of Polymerization, Volume 1: Vinyl Polymerization (Part I)* (G. E. Ham, ed.), Marcel Dekker, Inc., New York, 1969.

13. W. H. Stockmayer, "Distribution of Chain Lengths and Compositions in Copolymers," *J. Chem. Phys.*, *13*, 199 (1945).
14. F. W. Billmeyer, Jr., *Textbook of Polymer Science*, 3rd ed., Wiley, New York, 1984.
15. Richard N. Mead, Emulsion Copolymerization in Continuous Reactors, Ph.D. Thesis, Georgia Institute of Technology, 1987.

Bibliography

J. Brandrup and E. H. Immergut (eds.), *Polymer Handbook, Third Edition*, Wiley, New York, 1989.

D. C. Chappelear and R. H. Simon, "Polymerization Reaction Engineering," *Adv. Chem.*, *91*, 1–24 (1969).

E. B. Nauman, "Polymer Reaction Engineering," in *Chemical Reactor Design*, Wiley, New York, 1987.

W. H. Ray, and R. L. Laurence, "Polymerization Reaction Engineering," in *Chemical Reactor Theory: A Review* (Leo Lapidus and Neil R. Amundson, eds.), Prentice-Hall, Englewood Cliffs, NJ, 1977.

3

Polymerization Reaction Engineering

As the reader must realize by now, polymerizations involve highly complex reaction networks producing highly complex molecular structures. Conditions such as the locus of polymerization will have a profound effect on NACL, degree of branching, CCD, CSD, and other measures of molecular structure directly affecting the end-use properties. Because of this, it is often said that polymers are "products by process." By this is meant that the type of process (batch, semibatch, or continuous polymerization), as well as the processing conditions (recipe, temperature, etc.), will to a large degree define the structure of the resulting polymer and, hence, its end-use properties. This chapter will explore the effect of reactor type and operation on molecular structure. In the narrow sense, this is what is meant by *polymerization reaction engineering*. It is what differentiates the polymerization chemistry of the synthetic organic chemist from the polymerization reaction engineering of the chemical engineer.

3.1 Reactor Types

The application of the various idealized continuous reactors to polymerization systems will now be discussed. For a detailed discussion of reactor types and the assumptions inherent in their characterization, the reader is referred to a standard text on reactor design such as Hill [1].

Three types of reactors will be considered here: the batch (or semibatch) reactor, the plug flow reactor (PFR), and the continuous stirred tank reactor (CSTR). The concept of a batch reactor should be obvious to the reader. The ramifications of this type of reactor on polymer product will be discussed later.

The plug flow or tubular reactor is, as the names imply, a tube through which the reaction fluid travels in plug flow. The reactor may be a long, jacketed tube, or a coil immersed in a heat transfer fluid. Because the reaction fluid travels in plug flow, there is no axial mixing (no backmixing). Because the fluid must be in turbulent flow to achieve the plug flow condition, radial mixing is assumed to be perfect. Whereas these assumptions are never strictly correct, they offer a realistic (approximate) description of a great many tubular reactors. With these assumptions, each element of reaction mixture can be viewed as passing through the reactor without interacting with the elements before and after it. Because of this, the PFR is kinetically identical to the batch reactor.

The other common continuous polymerization reactor is the continuous stirred tank reactor. In this configuration, reactants are continuously pumped into a reactor while product is continuously removed. Because the reactor is assumed to be well-mixed, the product stream has a composition equal to that of the contents of the reactor.

To investigate the impact of reactor configuration on the polymer product to be made, one must understand the environment to which the reacting materials are exposed. Figure 3.1 contrasts the residence time distributions for the PFR and the CSTR. The ordinate of each of these plots shows the fraction of the outlet stream which has remained in the reactor for a time denoted by the abscissa. Because the PFR is, by definition, in plug flow, the residence time distribution consists of a narrow peak at the average residence time. Thus, all material leaving the PFR has remained in the reactor for the exact same time. On the other hand, each element of fluid has, during its residence in the reactor, experienced a varying environment in terms of monomer concentration, etc. In the CSTR, every element in the reactor has an equal chance of appearing in the outlet flow because the reactor is assumed to be well-mixed. Because of this, the residence time distribution take on the form shown in Fig. 3.1. Thus, some elements of reacting material will remain in the reactor for long periods, whereas others will leave the reactor almost immediately upon entering. On the other hand, in a CSTR at steady state, an element of fluid will experience the same concentration environment over the duration of its entire residence in the reactor. In summary, the PFR provides an identical residence time for each element of fluid but in a time-varying concentration environment, whereas the CSTR allows a variation in residence time for different elements of fluid but provides a constant concentration environment. This fundamental difference in reactor characteristics results in a significant influence of reactor type on polymer structure and mandates the use of one or the other reactor type for specific combinations of kinetics and desired polymer properties.

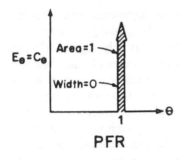

PFR

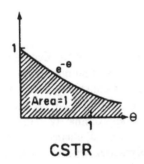

CSTR

Figure 3.1 Residence time distributions for the ideal PFR and CSTR. Residence time versus fraction of outlet stream having that residence time.(From Ref. 2 with permission.)

3.2 Batch/Semibatch Reactors

The most common polymerization reactor on a numerical basis is the batch kettle. (However, on a poundage basis, continuous reactors dominate, due mainly to the continuous polymerization of very high volume polymers such as polyolefins and polyesters.) Batch kettles may range in size from a 5-gal pilot plant kettle, to a 30,000-gal production kettle. They are generally constructed of stainless steel or, in cases where fouling of the walls with polymer is severe, may be glass-lined. They may be heavily instrumented or operated without controls. Removal of the heat of polymerization is accomplished by circulating coolant through a jacket or by refluxing monomer and solvent. Batch reactors have the advantage of flexibility to accommodate multiple products, but suffer from the disadvantage of batch-to-batch variability.

If all reactants are added at the beginning of the polymerization, the kettle is said to be operating in the *batch* mode. If a reactant (such as a comonomer) is added during the course of the polymerization, the kettle is said to be operating in the *semibatch* mode. If a small stream of initiator is added over time in a free radical polymerization (perhaps to control the rate of polymerization), the reactor is, strictly speaking, a semibatch process. Practically speaking, if the initiator flow is small, such that the total volume of the reactor is only slightly affected by its addition, the initiator stream is to be considered a control input, and the polymerization is often referred to as a batch process. Similarly, a condensation polymerization might be considered strictly semibatch in that condensation product is continuously removed to drive the reversible polymerization toward polymer rather than monomer. Once again, in practice, such a polymerization is considered to be batch. Batch/semibatch kettles have the economic advantage of flexibility for multiple products. On the other hand, batch-to-batch variations may be a problem.

3.2.1 Batch Reactors

Kinetically, the significance of the batch polymerizer lies in the fact that all reactants are added at the beginning of the polymerization, and nothing is removed until the end. Thus, the rate of polymerization changes constantly with time as the monomer and initiator concentrations decrease. In a step-growth polymerization, where the growth time of an individual chain is approximately kettle time, the effects of changing conditions are not critical because all chains will see the same (changing) environment. In a free radical polymerization, where the time of formation of a single chain is a small fraction of the kettle time, a batch reactor results in inhomogeneity because polymer chains are formed under very different conditions. This is especially significant in batch copolymerization, where polymer chains formed early in the reaction may contain a high percentage of the more reactive monomer, whereas those formed later will contain a high percentage of the remaining (less reactive) monomer.

In considering batch polymerization systems, the kinetic equations derived in Chapters 1 and 2 can be used directly. For anionic polymerization with chain transfer, the instantaneous degree of polymerization, given by Eq. (1.20) was shown to be a function of the monomer concentration:

$$\bar{x} = \frac{k_p M}{k_f B} \tag{1.20}$$

Thus, if chain transfer is significant, a polymer formed at the beginning of the polymerization (high monomer concentration) will have a higher average molecular weight than that formed late in the polymerization (at low monomer concentration). If, on the other hand, chain transfer is insignificant, Eqs. (2.46) and (2.47) show the polydispersity approaching unity at high conversion, making anionic polymerization a good method for manufacturing high-molecular-weight standards:

$$\mu_n = (1 + \tau) \approx \tau = \sigma_n^2 \tag{2.46}$$

$$D = 1 + \frac{\sigma_n^2}{\mu_n^2} = 1 + \frac{1}{\mu_n} \qquad (2.47)$$

For free radical polymerization, the instantaneous degree of polymerization is given by Eq. (1.48):

$$\bar{x} = \frac{k_p MP}{\frac{1}{2}k_{tc}P^2 + k_{td}P^2 + k_{fm}MP + k_{fs}PS + k_{fi}PT} \qquad (1.48)$$

Once again, the instantaneous degree of polymerization will vary over a batch polymerization cycle. Equations (2.79) and (2.82) show the polydispersity of the dead polymer assuming termination by disproportionation and combination, respectively:

$$D = \frac{\mu_w}{\mu_n} \approx 2 \qquad \text{(dead polymer termination by disproportionation)} \qquad (2.79)$$

$$D = \frac{\mu_w}{\mu_n} = \frac{2 + \alpha}{2} \approx 1.5 \qquad \text{(dead polymer termination by combination)} \qquad (2.82)$$

It must be noted, however, that Eqs. (2.79) and (2.82) were derived by assuming monomer and polymer concentrations constant in Eq. (2.75). Thus, Eqs. (2.79) and (2.82) represent limiting cases for the polydispersity which necessarily must be greater in a batch reactor.

For step-growth polymerization, Eq. (2.109) shows the NACL to be a function of extent of reaction only:

$$\mu_n = \frac{\lambda_1}{\lambda_0} = \frac{2 - \alpha}{1 - \alpha} = \frac{2 - p}{1 - p} \qquad (2.109)$$

Equation (2.111) shows that, under conditions of complete conversion, the polydispersity reduces to a minimum value of 2:

$$D = \frac{\mu_w}{\mu_n} = \frac{1 - \alpha}{(2 - \alpha)^2} + \frac{2}{(2 - \alpha)^2} \qquad (2.111)$$

Thus, batch polymerization is well-suited to step-growth polymerization because the reaction can be carried easily to a high extent of reaction to give a high molecular weight and a low polydispersity.

Design of batch reactors should be based on estimates of polymerization time from pilot plant data and/or simulation studies. If estimates of the charge, discharge, and servicing times associated with each batch are added to the actual polymerization time, the resultant cycle time can be used to determine the size and number of kettles necessary to produce a given name plate capacity. In scale-up, special care should be given to mixing and heat transfer. Both tend to degrade as the kettle size is increased. In designing for free radical polymerization, the heat transfer capability should be designed for what is known as the *exotherm*, or the point at which, due to the gel effect, the rate of polymerization is at a maximum. Due to the resultant inhomogeneities, incomplete mixing will result in a broadening of the property distributions describing the product.

Optimization of batch kettle operation will include considerations of kettle time versus conversion, kettle time versus monomer recovery cost, and the potential for variations in polymerization temperature within a batch to achieve desired product properties.

3.2.2 Semibatch Reactors

As noted previously, semibatch copolymerization is most often done in an attempt to maintain a reasonably constant copolymer composition when the comonomers are of widely varying reactivities [3]. Semibatching of initiator is often done to maintain temperature control in a heat transfer-limited kettle, and semibatch addition of an initiator or chain transfer agent may be used to maintain a desired MWD. Quantitative strategies for semibatching may be developed through empirical experimentation at the bench or pilot scale, or, if accurate mathematical models are available, classical trajectory optimization techniques may be used [4,5].

3.3 Plug Flow Reactors

In a plug flow reactor, each element of the reaction mixture can be viewed as an individual batch reactor. The batch time is the residence time in the tubular reactor, which is easily calculated as the total volume of the tube divided by the volumetric flow rate. Because no material enters or leaves the fluid element during the reaction time, all of the kinetic relationships derived thus far for the batch reactor are directly applicable to the plug flow reactor. The plug flow reactor, then, becomes the reactor of choice if it is desired to exploit the kinetic advantages of the batch reactor (high conversion, etc.) while enjoying the operational advantages of continuous processing (ease of operation, lack of batch-to-batch variability). This, in fact, has been done extensively in the production of high-volume step-growth polymers such as polyester and the nylons where the requirement for a high extent of reaction to obtain a high molecular weight is coupled with the requirement of continuous processing to obtain product uniformity and reduce production costs.

Tubular reactors (approximating plug flow characteristics) are applicable in high-volume polymerizations and exhibit excellent heat transfer capabilities because the surface-to-volume ratio is quite large for a tube with a high aspect ratio. On the other hand, they are particularly vulnerable to fouling because small amounts of polymer can plug the tube and shut down the entire process. Temperature programming to obtain specific polymer properties is possible by segregating the cooling jackets into separate zones along the length of the tube and controlling the temperature in each zone separately.

For viscous solutions, the assumptions of plug flow are not strictly valid. If the velocity profile is not flat, the polymer solution near the tube wall will move more slowly than that near the center of the tube. Because the slow-moving polymer near the wall remains in the reactor longer, it will polymerize to a higher conversion (or extent of reaction) than the bulk material. This higher conversion will then compound the viscosity problem. Studies on the effect of this deviation from plug flow in tubular polymerization have be done by Hamer and Ray [6,7].

3.4 Continuous Stirred Tank Reactors

The use of continuous stirred tank polymerization (in a single CSTR or train multiple CSTRs in series) may be warranted for high-volume products. The nature of the reactor system results in low processing costs, high throughput, and, in most cases, a highly uniform product. The fact that the polymerization rate is constant will contribute to product homogeneity, but may be overshadowed by inhomogeneity induced by the broad residence time distribution, particularly for emulsion free radical polymerizations. Large residence time CSTR systems are not particularly flexible and are, therefore, best suited to extended production runs of a small number of products. In low residence time CSTRs (as in olefin polymerization), grade changes can be made rapidly and low-volume products can be made effectively.

Because the rate of polymerization will vary with operating conditions, a CSTR system must be sized for the rate of polymerization at design conditions. Adequate heat transfer capability must be incorporated because most operating systems are heat transfer limited. In a CSTR train, overdesign of the cooling system maybe warranted because a large portion of the total heat of polymerization may be liberated in a single reactor in the train. The location of the exotherm will vary with operating conditions and product, so each CSTR should have sufficient heat transfer capability to accommodate this exotherm. For free radical polymerizations with a significant gel effect, the total reactor volume may be minimized by sizing each reactor independently as is shown in Hill [1] for autocatalytic reactors. The value of this is questionable because, as noted earlier, the location of the exotherm (maximum rate of polymerization) will vary with operating conditions and recipe. Instead, general practice is to use a train of equal-sized reactors to minimize design complexity.

In summary, the sizing of the reactors should be based on rate of polymerization data from pilot-scale investigations or (preferably) previous plant experience, whereas the choice of reactor type should be made based on kinetic considerations. The effect of the CSTR residence time distribution on each of the three common types of polymerization kinetics will now be discussed.

3.4.1 Anionic Polymerization

Following the general approach of Tirrell and Laurence [8], if one assumes rapid initiation and no termination or chain transfer, the monomer and total polymer balances for anionic polymerization are given:

$$\frac{dM}{dt} = Q_f M_f - QM - Vk_p PM, \qquad M(0) = M_0 \tag{2.1}$$

$$V\frac{dP}{dt} = Q_f P_f - QP, \qquad P(0) = P_0 \qquad P = \sum_{n=1}^{\infty} P_n \tag{2.2}$$

Balances on P_1 and P_n may be written as

$$V\frac{dP_1}{dt} = Q_f P_{1f} - Vk_p MP_1, \qquad P_1(0) = P_{10} \tag{3.1}$$

$$V\frac{dP_n}{dt} = Q_f P_{nf} - QP_n - Vk_p M(P_n - P_{n-1}), \qquad P_n(0) = 0, n \geq 2 \tag{3.2}$$

If the inflow and outflow terms are set equal and the feed is assumed to contain only P_1, the following steady-state balances may be obtained from Eqs. (2.1), (2.2), (3.1), and (3.2):

$$P = P_f = P_{1f} \tag{3.3}$$

$$Q(M_f - M) = Vk_p MP \tag{3.4}$$

$$Q(P_{1f} - P_1) = Vk_p MP_1 \tag{3.5}$$

$$QP_n = Vk_p M(P_n - P_{n-1}) \tag{3.6}$$

This system of difference equations may be solved by z-transforms to give

$$M = \frac{M_f}{(1 + \Theta k_p P_{1f})}, \qquad \theta = \frac{V}{Q} \tag{3.7}$$

$$P_n = \frac{P_{1f}}{(1 + \Theta k_p M)} \left[\frac{\Theta k_p M}{(1 + \Theta k_p M)} \right]^{n-1} = P_{1f}(1 - \emptyset)\emptyset^{n-1} \tag{3.8}$$

$$\mu_n = \frac{1}{1 - \emptyset} \tag{3.9}$$

$$D = 1 + \emptyset \tag{3.10}$$

$$\emptyset = \left[\frac{\Theta k_p M}{1 + \Theta k_p M} \right] \tag{3.11}$$

This is the Flory or Most Probable Distribution [9] and considerably broader ($D > 1$) than the Poisson distribution ($D = 1$) resulting from batch polymerization. The broader distribution is, of course, due to the broad residence time distribution of the CSTR. A train of CSTRs in series will give the system more of a PFR character, and the resulting polydispersity will be between that of the PFR and that of the single CSTR.

3.4.2 Free Radical Polymerization

Consider the free radical polymerization mechanism defined by Eqs. (1.33)–(1.41). Ignoring inhibition [although it may be easily accounted for via Eq. (1.47)] and considering CSTR solution polymerization, the proper mass balances may be written assuming constant reactor volume and isothermal operation as follows [10,11]. If q is the volumetric feed rate and the subscript f indicates feed conditions, balances over monomer and initiator yield:

$$\frac{V \, dM}{dt} = Q(M_f - M) - k_p PMV, \qquad M(0) = M_0 \tag{3.12}$$

$$\frac{dI}{dt} = Q(I_f - I) - k_d I, \qquad I(0) = I_0 \tag{3.13}$$

Additionally, Eqs. (1.43) and (1.44) may be written to describe the concentrations of primary radicals and total live chains. As before, the time derivatives of R and P are set to zero, and R is eliminated from the two equations to give Eq. (1.45):

$$P = \left[\frac{2 f k_d I}{k_{tc} + k_{td}} \right]^{1/2} \tag{1.45}$$

Equations (3.12), (3.13), and (1.45) define the conversion-time behavior of the reactor. Equation (1.45) does not account for the loss of live radicals by washout. However, Ray [12] has shown this to be insignificant.

Equation (1.48) defines the instantaneous degree of polymerization at any instant during the reaction. However, if complete information about the chain length distributions of the live and dead polymer is desired, balances must be made over each of these species. Balances over live chains of length 1 and n ($n > 1$), neglecting chain transfer to polymer and chain transfer agent, result in

$$\frac{V \, dP_1}{dt} = k_1 R M_1 V - k_p P_1 M_1 V + (k_{fs} S + k_{fm} M_1)(P - P_1)V$$

$$- (k_{tc} + k_{td}) P P_1 V - Q P_1, \qquad P_1(0) = 0 \qquad (3.14)$$

$$\frac{V \, dP_n}{dt} = k_p M_1 (P_{n-1} - P_n)V - (k_{fs} S + k_{fm} M_1) P_n V$$

$$- (k_{tc} + k_{td}) P P_n V - Q P_n, \qquad P_n(0) = 0 \quad n > 1 \qquad (3.15)$$

A balance over dead chains of length n ($n > 1$) can be written as

$$\frac{V \, dM_n}{dt} = (k_{fs} S + k_{fm} M_1) P_n V + k_{td} P_n P V + \frac{1}{2} V k_{tc} \sum_{m=1}^{n-1} P_m P_{n-m}$$

$$- Q M_n, \qquad M_n(0) = 0; \, n > 1 \qquad (3.16)$$

If the QSSA is made for P_1,

$$P_1 = (1 - \alpha)P \qquad (3.17)$$

where the probability of propagation is now given as

$$\alpha = \frac{k_p M}{k_p M + k_{fm} M + k_{fs} S + (k_{tc} + k_{td})P + 1/\Theta} \qquad (3.18)$$

where the reactor residence time is defined as

$$\Theta = \frac{V}{Q} \qquad (3.19)$$

Likewise, making the QSSA for P_n results in

$$P_n = \alpha P_{n-1} \qquad (3.20)$$

Solving Eqs. (3.17) and (3.20) by z-transform techniques as in Section 2.1 results in

$$P_n = (1 - \alpha)P\alpha^{n-1} \qquad (3.21)$$

Note that this once again is the Flory or "most probable" distribution.

Analysis of the dead polymer NCLD may be carried out in a manner analogous to that in Section 2.2 (for batch polymerization). The resulting equations for the time derivatives of the leading moments of the dead polymer NCLD can be written as

$$\frac{d(V\eta_0)}{dt} = (k_{fm}M + k_{td}P + k_{fs}S)PV\alpha + 0.5k_{tc}P^2V, \qquad \eta_0(0) = 0$$

$$\text{(3.22)}$$

$$\frac{d(V\eta_1)}{dt} = \frac{[(k_{fm}M + k_{td}P + k_{fs}S)(2\alpha - \alpha^2) + k_{tc}P]VP}{1 - \alpha},$$

$$\eta_1(0) = 0 \qquad\qquad\qquad\qquad\qquad\qquad\qquad \text{(3.23)}$$

$$\frac{d(V\eta_2)}{dt} =$$

$$= \frac{[(k_{fm}M + k_{td}P + k_{fs}S)(\alpha^3 - 3\alpha^2 + 4\alpha) + k_{tc}P(\alpha + 2)]VP}{(1 - \alpha)^2},$$

$$\eta_2(0) = 0 \qquad\qquad\qquad\qquad\qquad\qquad\qquad \text{(3.24)}$$

where the probability of propagation is given by Eq. (3.18) rather than Eq. (2.54).

The moments of the live and dead NCLD may now be evaluated. The moments of the live NCLD are identical to those for batch polymerization, but with Eq. (3.18) defining the probability of propagation:

$$\lambda_0 = P \qquad\qquad\qquad\qquad\qquad\qquad\qquad\qquad \text{(3.25)}$$

$$\lambda_1 = \frac{P}{1 - \alpha} \qquad\qquad\qquad\qquad\qquad\qquad \text{(3.26)}$$

$$\lambda_2 = \frac{P(1 + \alpha)}{1 - \alpha} \qquad\qquad\qquad\qquad\qquad \text{(3.27)}$$

Using Eqs. (2.10), (2.13), and (1.5), the NACL, WACL, and polydispersity can be calculated from the moments

$$\mu_n = \frac{\lambda_1}{\lambda_0} = \frac{1}{1 - \alpha} \qquad \text{(live polymer)} \qquad\qquad \text{(3.28)}$$

$$\mu_w = \frac{\lambda_2}{\lambda_1} = \frac{1 + \alpha}{1 - \alpha} \qquad \text{(live polymer)} \qquad\qquad \text{(3.29)}$$

$$D = \frac{m_w}{m_n} = \frac{\mu_w}{\mu_n} = 1 + \alpha \qquad \text{(live polymer)} \qquad \text{(3.30)}$$

For long chains, the probability of propagation approaches unity so that

$$D = 1 + \alpha \approx 2 \qquad \text{(live polymer)} \tag{3.31}$$

To derive the NCLD for the dead polymer, Eqs. (3.22)–(3.24) must be solved for the steady state values of the moments.

The NACL and WACL of the dead polymer distribution can then be calculated as follows:

$$\mu_n = \frac{\eta_1}{\eta_0} = \frac{(k_{fm}M_1 + k_{td}P + k_{fs}S)(2\alpha - \alpha^2) + k_{tc}P}{(1 - \alpha)[(k_{fm}M_1 + k_{td}P + k_{fs}S)\alpha + 0.5k_{tc}P]} \tag{3.32}$$

$$\mu_w = \frac{\eta_2}{\eta_1} = \frac{(k_{fm}M_1 + k_{td}P + k_{fs}S)(\alpha^3 - 3\alpha^2 + 4\alpha) + k_{tc}P(\alpha + 2)}{(1 - \alpha)[(k_{fm}M_1 + k_{td}P + k_{fs}S)(2\alpha - \alpha^2) + k_{tc}P]} \tag{3.33}$$

Substituting Eqs. (3.17) and (3.21) into Eq. (3.16) results in

$$\frac{dM_n}{dt} = [k_{fs}S + k_{fm}M_1 + k_{td}P](1 - \alpha)P\alpha^{n-1}$$

$$+ \frac{1}{2} k_{tc}P^2(1 - \alpha)^2 \alpha^{n-2}(n - 1) - \frac{1}{\Theta} M_n, \qquad M_n(0) = 0, \quad n > 1 \tag{3.34}$$

Assuming steady-state conditions and solving for M_n,

$$M_n(t) = \Theta \begin{cases} [k_{fs}S + k_{fm}M_1 + k_{td}P](1 - \alpha)P\alpha^{n-1} \\ + 1/2 \, k_{tc}P^2(1 - \alpha)^2 \alpha^{n-2}(n - 1), \end{cases} \quad M_n(0) = 0, \, n > 1 \tag{3.35}$$

Two special cases may now be considered. If the long chain approximation is made and there is no termination by combination,

$$M_n(t) = \eta_0(1 - \alpha)\alpha^{n-2} \tag{3.36}$$

The power of $n - 2$ comes about because in this treatment we have excluded M_1 from consideration as a polymer. If M_1 (here meant to indicate a polymer of length unity formed by a termination or transfer reaction from P_1, and not monomer) is included in the normalization, the distribution may be written as

$$M_n(t) = M_{tot}(1 - \alpha)\alpha^{n-1} \tag{3.37}$$

where M_{tot} is taken to mean all polymer of length 1 or greater. This is again the Flory distribution and is exactly the distribution defined by Eq. (3.21). Thus, for termination by disproportionation only, the dead polymer will have the same distribution as the live polymer. Assuming the probability of propagation to be unity, the polydispersity of the dead polymer may be calculated as the ratio Eq. (3.33) divided by Eq. (3.32):

$$D = \frac{\mu_w}{\mu_n} \approx 2 \quad \text{(dead polymer)} \tag{3.38}$$

From Eqs. (3.28) and (3.32), it may be seen that

$$\frac{\mu_n(\text{dead})}{\mu_n(\text{live})} = 2 - \alpha \approx 1 \tag{3.39}$$

If, instead, termination by combination is assumed to be the only mode of chain termination (no chain transfer to solvent or monomer, and no termination by disproportionation),

$$M_n(t) = \eta_0(n - 1)(1 - \alpha)^2 \alpha^{n-2} \quad \text{(dead polymer)} \tag{3.40}$$

Because no M_1 chains can be formed by combination, M_{tot} is equal to the zeroth moment of the dead polymer distribution, and Eq. (3.40) is equivalent to both Eqs. (3.36) and (3.37). The polydispersity, for long chains, is

$$D = \frac{\mu_w}{\mu_n} = \frac{2 + \alpha}{2} \approx 1.5 \quad \text{(dead polymer)} \tag{3.41}$$

From Eqs. (3.28) and (3.32), it may be seen that

$$\frac{\mu_n(\text{dead})}{\mu_n(\text{live})} = 2 \tag{3.42}$$

and the dead chains have twice the NACL of the live chains.

Thus, under ideal conditions, the polydispersity will range from 1.5 for termination by combination and no chain transfer, to 2.0 for termination by disproportionation. These are the same results obtained as a *limiting case* for the batch reactor. Thus, the CSTR will give a narrower dead polymer NCLD because it is possible to maintain a constant reaction environment at steady state. Note, also, that the effect of residence time on the NCLD is felt through the value of the probability of propagation [Eq. (3.18)]. In the case of free radical polymerization, the effect of the constancy of the reaction environment in the CSTR results in reduced polydispersity when compared with batch polymerization. The effect of the residence time distribution on the polydispersity is negligible because the lifetime of a single radical is far less than the average residence time.

Likewise, for a copolymerization in a CSTR at steady state, the constancy of the ratio of comonomer concentrations will result in a narrow copolymer composition distribution. The copolymer composition for various comonomer feed concentration ratios is shown in Fig. 2.5 for the CSTR copolymerization of styrene and methyl acrylate.

3.4.3 Step-Growth Polymerization

As shown in Chapter 2, step-growth polymerization kinetics can be represented by Eq. (2.92):

$$A-B + A-B \rightleftharpoons A-B-A-B + W \tag{2.92}$$

An analysis of the MWD during step-growth polymerization in a CSTR at steady state can be begun by making the assumption of irreversible polymerization (due, perhaps, to the continuous removal of the condensation product, W) and writing mass balances on the growing chains:

$$P_{1f} - P_1 = k_p \Theta P_1 P \tag{3.43}$$

$$P_{nf} - P_n = \frac{1}{2} k_p \Theta \sum_{r=1}^{n-1} P_r P_{n-r} + k_p \Theta P_n P, \quad n > 1 \tag{3.44}$$

$$P = \sum_{n=2}^{\infty} P_n \tag{3.45}$$

Assuming the feed consists of P_1 only, solution of Eqs. (3.43) through (3.45) by z-transform techniques (or by generating function techniques, as in Ref. 8) indicates that the NCLD is a binomial distribution with the following leading moments:

$$\lambda_0 = \frac{-1 + [1 + 2k_p \Theta P_{1f}]^{1/2}}{k_p \Theta} \tag{3.46}$$

$$\lambda_1 = P_{1f} \tag{3.47}$$

$$\lambda_2 = P_{1f}(1 + k_p \Theta P_{1f}) \tag{3.48}$$

The NACL, WACL, and polydispersity can then be calculated as

$$\mu_n = \frac{k_p \Theta P_{1f}}{[-1 + (1 + 2k_p \Theta P_{1f})^{1/2}]} \tag{3.49}$$

$$\mu_w = 1 + k_p \Theta P_{1f} \tag{3.50}$$

$$D = \frac{(1 + k_p \Theta P_{1f})[-1 + (1 + 2k_p \Theta P_{1f})^{1/2}]}{k_p \Theta P_{1f}} \tag{3.51}$$

The extent of reaction in a CSTR is defined as

$$p = \frac{P_{1f} - P}{P_{1f}} = \frac{1}{2} \frac{k_p \Theta P^2}{P_{1f}} \tag{3.52}$$

Equations (3.49)–(3.51) can be rewritten in terms of the extent of reaction as

$$\mu_n = \frac{1}{1 - p} \tag{3.53}$$

$$\mu_w = 1 + \frac{2p}{(1-p)^2} = \frac{1 + p^2}{(1 - p)^2} \tag{3.54}$$

$$D = \frac{1 + p^2}{1 + p} \tag{3.55}$$

A number of important points can be made from these last equations. As with step-growth polymerization in a batch reactor, the extent of reaction must be almost unity before a large NACL can be achieved. High conversion products usually cannot be made economically in a single CSTR because this would require the CSTR to operate at almost complete conversion, and would require a reactor of very large volume. In addition, note that as the extent of reaction approaches unity (giving a high-molecular-weight product), the polydispersity increases without bound. For both these reasons, step-growth polymerization is rarely carried out in a single CSTR. A series of CSTRs or, preferably, a PFR is preferred if step-growth polymerization is to be carried out continuously.

In summary, because the lifetime of a growing polymer chain is equal to its residence time in the reactor, the effect of the residence time distribution causes extreme broadening of the molecular weight distribution during step-growth polymerization in a CSTR. The constancy of the polymerization environment, which acted to narrow the distribution in free radical polymerization, has an insignificant effect in step-growth polymerization.

3.4.4 Reactor Dynamics

As we have seen, the use of a CSTR or CSTR train for polymerization reactions may be justified in some cases by kinetic considerations. However, before implementing CSTR polymerization, the engineer should be aware of the unique dynamics associated with exothermic and/or autocatalytic reactions in a CSTR.

Consider an irreversible first-order exothermic reaction in a CSTR. The rate of thermal energy release by reaction can be plotted versus temperature as shown by the curve Q_g in Fig. 3.2. At low temperatures, the reaction rate is low and the slope of Q_g is slight. At high temperatures, the reactor is operating at a high level of conversion (low reactant concentration), and additional increases in

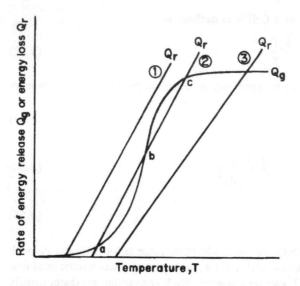

Figure 3.2 Heat balance multiplicity during exothermic reaction in a CSTR. (From Ref. 1 with permission.)

temperature result in a negligible increase in reaction rate and heat evolution. If the reactor is jacketed, the rate of heat removal (for fixed jacket temperature) is linear with reaction temperature. Thus, depending on operating conditions, the rate of heat removal may be represented by the various heat removal lines marked Q_r in Fig. 3.2. Because, at steady state, the rate of heat generation must equal the rate of heat removal, steady-state conditions can exist only at the intersection of the Q_g and Q_r curves. Depending on operating conditions (the slope and position of the Q_r line), there may be one or three steady states. In the case of three steady states, it may be seen easily that the upper and lower steady states are stable because perturbations in temperature will result in the system returning to its original position when the perturbation is removed. The middle steady state, however, can be seen to be unstable because any perturbation will drive the system away from the middle steady state and toward the upper or lower steady state (depending on the direction of the perturbation). This type of heat balance multiplicity is common in CSTR polymerization due to the highly exothermic nature of polymerization reactions.

A similar phenomenon can be observed in isothermal free radical polymerization in a CSTR. Figure 3.3 shows the rate of polymerization versus monomer conversion for the free radical solution polymerization of methyl methacrylate. Unlike a more common reaction in which the rate of reaction falls monotonically with conversion, the rate of reaction rises with conversion due to the onset of

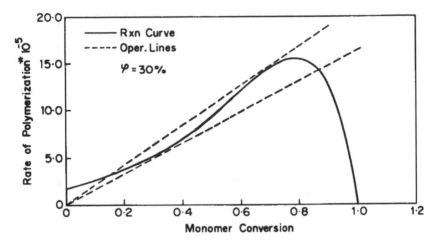

Figure 3.3 Mass balance multiplicity during free radical solution polymerization in a CSTR. (From Ref. 13 with permission.)

the gel effect. Thus, the system can be thought of as autocatalytic. At high conversions, the polymerization becomes monomer-starved and the rate of polymerization falls to zero. At a fixed residence time, there must be a specific rate of polymerization to produce a given monomer conversion. Thus, the mass balance is represented by the dotted lines in Fig. 3.3. The slope of the mass balance line will vary with operating conditions, but it will always pass through the origin because at zero reaction rate the monomer conversion is zero. Inspection of Fig. 3.3 will reveal that for mass balances (operating lines) with slopes between the two dotted lines, three steady states will exist because an intersection of the reaction rate curve and the operating line define a steady state. This may be better seen by referring to Fig. 3.4, where monomer conversion has been plotted versus reactor residence time. A similar plot will result from the heat balance multiplicity. It may be seen that, over a range of residence times, three values of monomer conversion are possible. As before, the upper and lower steady states are usually stable, whereas the middle steady state is not. As will seen in Fig. 4.2, a similar phenomenon is seen in emulsion polymerization for similar reasons.

Steady-state multiplicity can be an operational problem for a number of reasons. If one wishes to operate at an intermediate level of monomer conversion, one may be forced to operate in the unstable region, relying on closed-loop control to stabilize the operating point. This is tricky at best. Additionally, the steady state (upper or lower) to which the system goes on start-up will depend on how the start-up is effected. A careful start-up policy may be needed to assure that the system arrives at the desired steady state. Finally, large upsets in the process

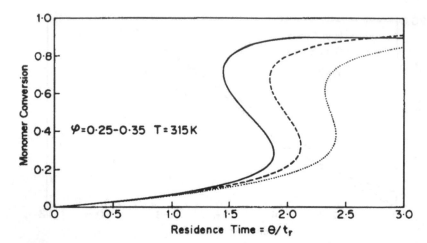

Figure 3.4 Multiplicity during isothermal CSTR solution polymerization effect of solvent volume fraction (θ). Dotted line: $\theta = 0.25$; dashed line: $\theta = 0.30$; solid line: $\theta = 0.35$. (From Ref. 13 with permission.)

may move the system from the desired (upper or lower) steady state to the other (stable) steady state. A system designed to operate at the lower steady state may not have the heat transfer capacity to operate safely at the upper steady state. A system designed to operate at the upper steady state will be operating way below design product yield at the lower steady state. Additionally, the product quality (MWD, CCD, etc.) will be different for the two operating points. The polymerization reactor designer should be aware of the potential for multiplicity and, if possible, design the system to operate outside the region of multiplicity.

CSTR polymerization reactors can also be subject to oscillatory behavior. A nonisothermal CSTR free radical polymerization can exhibit a damped oscillatory approach to a steady-state, unstable (growing) oscillations upon disturbance, and stable (limit cycle) oscillations in which the system never reaches steady state and never goes unstable, but continues to oscillate with a fixed period and amplitude. These phenomena are more commonly observed in emulsion polymerization, and examples of these behaviors are given in Chapter 4.

Damped oscillations will result in lost productivity because the product during these transients may be off-quality. Unstable oscillations will, of course, preclude continued operation. Limit cycle oscillations, while not unstable, will result in a product whose quality (MWD, CCD, etc.) varies with time in a cyclic fashion. In most cases, this is undesirable. As in the case of multiplicity, the polymerization reactor designer must be aware of the potential for oscillatory phenomena and should attempt to specify operating conditions at which these phenomena do not exist.

3.5 Reactor Selection

We have investigated the various types of chemical reactors and their effect on the properties (MWD, CCD, etc.) of the polymer produced. By way of summary, we will now review the selection of reactor type based on polymerization kinetics as well as practical considerations. Because polymers are "products by process," it is not surprising that very specific conclusions can be drawn regarding the reactor selection for a required combination of polymerization capacity and product quality.

In terms of practical (operational) considerations, batch reactors are well-suited to low-volume products and to products for which there are numerous grades. Each batch may be made according to its own recipe and operating conditions without the waste incurred when a continuous reactor is shut down and restarted. These reactors, however, may suffer from batch-to-batch variations in product properties. Continuous reactors, on the other hand, are best suited to long campaigns of a single high-volume product. They offer low operating costs and consistent products, but at the cost of diminished flexibility.

Table 3.1 (adapted from Gerrens [14,15]) summarizes the effects of reactor type on MWD as determined by the analyses in Chapters 2 and 3. Following Gerrens, polymerization kinetics have been divided into three categories: monomer linkage (addition polymerization) with termination (as in free radical polymerization); monomer linkage without termination (as in anionic polymerization); and polymer linkage (step-growth polymerization). Table 3.1 lists three categories of reactors: the batch reactor (BR) or PFR (because the kinetics are identical to those of a batch reactor); the homogeneous CSTR (HCSTR); and the segregated CSTR (SCSTR). The SCSTR, while not treated in the analysis to this point, is included to indicate the effects of less than perfect mixing. A practical example of an SCSTR is suspension polymerization in a CSTR (see Chapter 4) in which the suspension beads are well-mixed within the reactor, and within themselves, but in which there is no exchange of material between the various beads.

For a polymerization featuring monomer linkage with termination, the narrowest MWD is developed in the HCSTR. In this case, the constancy of the reaction environment (at steady state) dominates over the distribution of residence times in the reactor. This is due to the fact that the lifetime of a single live polymer chain is far less than the average reactor residence time. Due to the constantly changing monomer concentration in a BR, the MWD will be wider for a batch reactor. The distribution will be widest for the SCSTR where the effects of reaction environment and residence time variation combine to cause a broadening of the MWD. Copolymerization with the object of producing a narrow CCD is best carried out in an HCSTR because the ratio of comonomer concentrations stays constant at steady state.

Table 3.1 Molecular weight distribution versus reactor type.

Reactor	Monomer linkage[a]			Polymer linkage
	Termination	No termination		
Batch or PFR	(1A)[b]	(2A)		(3A)[c]
	Wider than Flory	Poisson		Flory
	($D > 1.5$–2.0)	($D = 1.0$)		($D = 2.0$)
Homogeneous	(1B)	(2B)		(3B)
CSTR	Flory	Flory		Wider than Flory
	($D = 1.5$–2.0)	($D > 1.0$)		($D > 2.0$)
Segregated	(1C)	(2C)		(3C)
CSTR	Wider than	Between		Between
	(1A)	(2A) and (2B)		(3A) and (3B)

[a]An example of monomer linkage with termination is free radical polymerization. An example of monomer linkage without termination is anionic polymerization.
[b](1A): $D = 1.5$ for termination by combination; $D = 2.0$ for termination by disproportionation; $1.5 < D < 2.0$ for mixed termination.
[c](3A), (3B), and (3C) assume an extent of reaction of 0.99+.

For monomer linkage without termination, the narrowest distribution occurs in the BR because all chains are growing throughout the reaction and the effect of constant reaction environment is moot. The HCSTR will exhibit a broader distribution due to the effect of variations in residence times among the growing chains. The SCSTR will have an MWD whose breadth is intermediate between the BR and HCSTR because the segregation (batch character) will cause a narrowing of the MWD, whereas the residence time variations among the segregated beads will cause a broadening.

For polymer linkage, the narrowest distribution is found in the BR for much the same reasons as in monomer linkage without termination. The HCSTR causes a broadening of the MWD because the lifetime of each growing chain is equal to its residence time in the reactor, and the distribution of residence times broadens the MWD. In fact, as was shown, as the extent of reaction approaches unity (necessary to produce a high-molecular-weight product), the polydispersity increases without bound. Thus, the HCSTR is never recommended for polymer linkage polymerizations. The SCSTR develops an MWD intermediate between the BR and the HCSTR because, once again, the segregation narrows the distribution, whereas the residence time distribution broadens it.

Of course, it is not always necessary, or even desirable, to produce a monodisperse MWD. A certain broadening may be designed into the product by the choice of reactor type; or, the engineer (chemist) may choose to produce a broad or bimodal MWD by blending two or more narrowly distributed products. As with most aspects of polymerization reaction engineering, the possibilities are endless.

Reactor selection for heterogeneous polymerizations may involve additional considerations and is covered in Chapter 4.

References

1. Charles G. Hill, Jr., *Chemical Engineering Kinetics and Reactor Design*, Wiley, New York, 1977.
2. Octave Levenspiel, *Chemical Reaction Engineering*, 2nd ed., Wiley, New York, 1972.
3. M. Tirrell and K. Gromley, "Composition Control of Batch Copolymerization Reactors," *Chem. Eng. Sci.*, 36, 367–375 (1981).
4. W. H. Ray, *Advanced Process Control*, McGraw-Hill, New York, 1981.
5. J. A. Hicks, A. Mohan, and W. H. Ray, "The Optimal Control of Polymerization Reactors," *Can. J. Chem. Eng.*, 47, 590–597 (1969).
6. J. Hamer, and W. H. Ray, "Continuous Tubular Polymerization Reactors–I. A Detailed Model," *Chem. Eng. Sci.*, 41, 3083–3093 (1986).
7. J. Hamer and W. H. Ray, "Continuous Tubular Polymerization Reactors–II. Studies of Vinyl Acetate Polymerization," *Chem. Eng. Sci.*, 41, 3095–3100 (1986).
8. M. V. Tirrell and R. L. Laurence, "Polymerization Reaction Engineering," Notes for a short course at the University of Minnesota Polymerization and Polymer Process Engineering Center, Minneapolis, MN, 1983.
9. J. A. Biesenberger and D. H. Sebastian, *Principles of Polymerization Engineering*, Wiley, New York, 1983.
10. A. D. Schmidt and W. H. Ray, "The Dynamic Behavior of Continuous Polymerization Reactors–I," *Chem. Eng. Sci.*, 36, 1401–1410 (1981).
11. B. M. Tanner, A. K. Adebekun, and F. J. Schork, "Feedback Control of Molecular Weight Distribution During Continuous Polymerization," *Polym. Proc. Eng.*, 5, 75–118 (1987).
12. W. H. Ray, "The Quasi-Steady-State Approximation in Continuous Stirred Tank Reactors," *Can. J. Chem. Eng.*, 47, 503–508 (1969).
13. K. M. Kwalik, "Bifurcation Characteristics in Closed-Loop Polymerization Reactors, " Ph.D. Thesis, Georgia Institute of Technology, 1988.
14. H. Gerrens, "How to Select Polymerization Reactors–Part I," *Chemtech*, 380–383, (June 1982).
15. H. Gerrens, "How to Select Polymerization Reactors–Part II," *Chemtech*, 434–443 (July 1982).

4

Heterogeneous Polymerization

Previously we have considered only polymerization in a homogeneous medium, either bulk or solution polymerizations. In some cases, it may be advantageous to employ a heterogeneous medium. In this chapter, we will consider the three most common types of heterogeneous polymerization: suspension, emulsion, and coordination polymerization.

Before considering these polymerization systems, we will note that if, during bulk polymerization, a polymer is insoluble in its monomer, the monomer will polymerize in the continuous phase, but upon reaching a certain chain length will precipitate from the monomer. In the absence of surface active agents (surfactants), the polymer will form a slurry phase which can be separated from the continuous phase by filtration. In the presence of surfactants, the polymer will form polymer particles which are colloidally stable, and the process will resemble, at least superficially, an emulsion polymerization. Such processes are known for acrylonitrile and vinyl chloride.

4.1 Suspension Polymerization

Both suspension and emulsion polymerization are carried out primarily with vinyl monomers employing the free radical mechanism as discussed in Chapter 1.

4.1.1 Process

Free radical polymerization can be carried out in bulk or solution, as well as in heterogeneous systems. In bulk polymerization, the viscosity of the monomer/polymer mixture rises very rapidly with monomer conversion. If substantial conversion is desired, the polymerization medium soon becomes too viscous for the effective removal of the heat of polymerization. This results in nonisothermal polymerization, uneven temperature distribution, and possible temperature runaway. In addition, unless polymerization and casting are done simultaneously, the viscosity of the product may make removal of the product from the reactor impossible.

The introduction of a third component which is a good solvent for both the monomer and polymer can be used to reduce the viscosity of the reaction mixture (solution polymerization). However, the cost of solvent recovery, the reduction in the rate of polymerization due to the dilution of the monomer, and the fact that large quantities of solvent are required to maintain fluidity at high conversion combine to reduce the economic viability of the solution polymerization process. Nevertheless, it is used in a number of major commercial applications.

If, instead, the monomer is dispersed in water prior to polymerization, the viscosity of the resulting suspension will remain essentially constant over the course of the polymerization. In addition, the product can be readily separated from the aqueous phase (via filtration or decantation) in the form of macroscopic particles or beads which can be easily packaged and/or transported. Heat transfer is facilitated by the presence of the continuous aqueous phase. This process is known as suspension or bead polymerization. An oil-soluble initiator is used. Blocking agents such as clays or talcs are used to prevent particle agglomeration. Small quantities of nonionic surfactants (polyvinyl alcohol, etc.) may be used to impart particle stability and to disperse the blocking agent. Viscosity enhancers such as carboxymethyl cellulose may be used to inhibit particle settling.

The locus of polymerization is the monomer/polymer beads. Due to the large size of the beads (0.1–1.0 mm), such systems are suspensions rather than emulsions or stable dispersions. The particles must be kept suspended by agitation throughout the course of the polymerization. The suspension polymerization process is described in detail by Trommsdorff and Schildknecht [1].

4.1.2 Kinetics

The diameter of a monomer droplet or bead in a suspension polymerization is quite large in comparison with the molecular dimensions on which the polymerization reactions take place. The initiators used (benzoyl peroxide, AIBN, etc.) are oil-soluble (completely miscible with the monomer phase) and only slightly soluble in the aqueous phase. Thus, the locus of polymerization is the monomer/polymer beads. Due to the large size of the beads, each acts as a separate batch reactor displaying bulk polymerization kinetics. This rate of polymerization may be calculated from Eq. (1.46) where the concentrations are those within the bead. Likewise, the NACL is given by Eq. (1.48). It will be noted from the above that both the rate of polymerization and the degree of polymerization are independent of the particle (bead) size and particle size distribution. The capacity of the suspension kettle, then, is a function of the monomer/water ratio and the initiator type and concentration, and not of the bead size or the parameters affecting bead size (agitation, suspending agent type and concentration, etc.).

4.2 Emulsion Polymerization

Since its first commercialization in the early 1930s, emulsion polymerization has grown to become one of the major processes for the production of synthetic polymeric materials. The loss of supply of natural rubber latex during World War II triggered a large research program aimed at producing styrene-butadiene rubber (SBR) by emulsion polymerization. Following the war, the development of water-based paint widened the use of emulsion polymerization into acrylic resins as well as synthetic elastomers. Today, a large variety of polymers is produced by emulsion polymerization including synthetic elastomers, bulk plastics, and plastic and elastomeric lattices for coatings. While the industry is still dominated by batch reactors, continuous processes are becoming more widespread due to their economic and processing advantages.

4.2.1 Process

The process of emulsion polymerization is, in essence, an attempt by man to imitate the production of natural rubber (polyisoprene) latex by the rubber tree (*Havea braziliensis*). Natural rubber latex is an aqueous dispersion of polyisoprene particles stabilized by proteinaceous surfactants. By analogy, an emulsion polymerized polymer latex is an aqueous dispersion of polymer particles (SBR, etc.) stabilized by a fatty acid soap or synthetic surfactant.

The polymer latex is formed by emulsifying a monomer (usually relatively insoluble in water) into an aqueous surfactant solution. By applying shear to the mixture, an emulsion is formed in which the monomer is dispersed into droplets of typically 10 μ diameter. The droplets are stabilized by a monolayer of surfactant at the monomer–water interface. Additional surfactant above the critical micelle concentration (CMC) is contained in micelles or aggregates of typically 50–100 surfactant molecules oriented so that the hydrophobic ends of the surfactant molecules are toward the center of the micelle while the hydrophilic ends extend out into the aqueous phase. The polymerization is initiated by adding a free radical initiator (usually water-soluble) and, as the polymerization proceeds, polymer particles stabilized by a surfactant layer are formed. Figure 4.1 shows

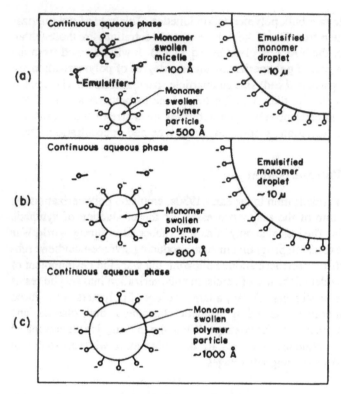

Figure 4.1 Smith–Ewart description of emulsion polymerization. (a) Interval I; (b) interval II; (c) interval III. (From Ref. 5 with permission.)

the main elements of emulsion polymerization: the continuous aqueous phase, monomer droplets, micelles, and polymer particles.

The emulsion polymerization process has significant advantages over other polymerization processes. The viscosity of the emulsion remains low even at high monomer conversion. Heat removal, a major consideration because most polymerizations are highly exothermic, is facilitated by the low viscosity and high heat capacity of the aqueous phase. For applications requiring a finished product in the form of a latex (i.e., latex paint), the polymerized emulsion may be used directly. Due to the nature of free radical polymerization kinetics, it is actually possible to attain a higher rate of polymerization in an emulsion system than in bulk polymerization.

4.2.2 Kinetics

Emulsion polymerization is primarily carried out with vinyl monomers via the free radical mechanism. The original theory of the mechanism of emulsion polymerization comes from Harkins [2–4]. With some modifications, Harkins' theory is the basis of the current understanding of the mechanism of emulsion polymerization.

Harkins' theory may be summarized as follows:

a. The initiator decomposes in the aqueous phase into free radicals which are captured by emulsifier micelles.
b. Each free radical begins the polymerization of solubilized monomer within the micelle. Once a free radical has entered a micelle, it becomes a growing polymer particle.
c. Additional monomer diffuses from the monomer droplets, across the aqueous phase and into the polymer particles where it sustains the growth of the particle by polymerization. As the particle grows, it adsorbs surfactant molecules from the micelles to stabilize its increasing surface area.
d. Micelles disappear by absorbing radicals to become polymer particles or by being depleted of their soap molecules by adsorption on growing polymer particles. Once all micelles disappear, no new particles are initiated.
e. Once initiated, a particle continues to grow until a second radical enters the particles and terminates the polymerization. A third radical can reinitiate polymerization and so on until all of the monomer is converted to polymer.

The concept of the polymer particle as the reaction locus is maintained today, although in some cases micellar initiation is not considered to be the only, or even the primary, mechanism of initiation.

A second initiation mechanism has been proposed which accounts for the experimentally observable phenomenon of particle formation in the absence of micelles. Fitch and Tsai [6] have given a development of the mechanism of homogeneous nucleation. In this mechanism, radicals in the aqueous phase initiate polymerization of monomer dissolved in the aqueous phase. Once the oligomers thus formed reach a certain degree of polymerization, they are no longer soluble and precipitate out of the aqueous phase, adsorb soluble surfactant, and nucleate new polymer particles. This is thought to be the primary mechanism of initiation with highly water-soluble monomers and a secondary source of initiation with less water-soluble monomers. Harkins' theory requires that each radical captured by a polymer particle either initiate polymerization or terminate the polymerization initiated by a previous radical. The possibility of radical desorption from the polymer particle is not considered. A mechanism of radical desorption and reabsorption has been postulated by a number of authors [7,8]. This is a significant phenomenon, especially in emulsion polymerization of the more water-soluble monomers such as vinyl acetate.

The characterization of the polymer–water interface at the surface of the polymer particle is, perhaps, the key to the understanding of emulsion polymerization. The values of surfactant adsorption, interfacial tension, and surface charge distribution can have profound effects on the rate of particle coalescence, radical adsorption and desorption, and rate of polymerization.

Surfactant adsorption on the surface of the polymer particle is usually considered to be an equilibrium process with respect to the much slower processes of particle growth and rate of polymerization. The surfactant isotherm may be affected by a number of factors. The ionic strength of the aqueous phase, by shifting the CMC of the surfactant system, will affect the surface coverage because the adsorbed surfactant is in equilibrium with the aqueous phase dissolved and a micellar surfactant. The saturation adsorption of surfactant is a function of the polarity of the monomer. For polar monomers, the particle–water interfacial tension is low and, consequently, the saturation or maximum surfactant coverage is reduced.

The surface coverage of the particles directly affects the course of the polymerization. Insufficiently protected particles are subject to flocculation that may change the particle size distribution and, hence, the reaction progress, or that may result in gross coagulation of the latex. It has also been proposed that, in systems where homogeneous nucleation is significant, the surface coverage (the amount and type of surfactant) strongly influences the rate of diffusion of oligomeric free radicals into the particles [9]. The Harkins theory assumes that the polymer particles are stabilized solely by adsorbed surfactant. In reality, experimental evidence has been accumulated which indicates that particles may be stabilized by polymer chain end groups which remain on the particle surface and, in the case of polar monomers, by polar groups on the polymer chains [10].

Interfacial tension between the polymer particle and the aqueous phase is a direct function of monomer polarity. A nonpolar monomer such as styrene results in a high interfacial tension. Conversely, a highly polar monomer such as vinyl chloride results in a low interfacial tension due to the orientation of the polar groups near the interface. The interfacial tension affects the particle in several ways. As discussed previously, low interfacial tension due to a polar monomer will tend to stabilize the particle. On the other hand, low interfacial tension results in low equilibrium surfactant adsorption and subsequent possible particle instability [11]. The most important effect of interfacial tension is on the equilibrium swelling of the polymer particle by the monomer. Equilibrium monomer concentration is controlled by the balance of the free energy of solution of the polymer in the monomer and the surface free energy of the particle. As such, the equilibrium swelling is a function of monomer polarity and particle size. Thus, a polar monomer will tend to swell the polymer particle more than a nonpolar monomer, resulting in a higher rate of polymerization due to a higher monomer concentration at the locus of polymerization.

Based on Harkins' picture of the process, Smith and Ewart [12–14] developed a simple model of batch emulsion polymerization. Their model may be divided into three intervals.

Interval I. Polymer particles are formed by the radical capture by micelles until all micelles disappear by radical capture or by dissolution and adsorption on existing particles.

Interval II. After the micelles disappear, no new particles are formed. Existing particles grow by chain propagation. An additional monomer is fed to the reaction locus by diffusion from the monomer droplets until these are consumed.

Interval III. From the point at which monomer droplets disappear, polymerization continues to full conversion by polymerization of the monomer swelling the polymer particles.

Based on this picture, Smith and Ewart developed the following equation for the number of particles at the end of Interval I (N_c):

$$N_c = k_N \left(\frac{\rho_w}{\mu} \right)^{0.4} (S_0)^{0.6} \qquad (4.1)$$

where

k_N = 0.53 if all radicals enter micelles and generate particles;
k_N = 0.37 if radicals enter both micelles and polymer particles competitively based on the relative surface areas;
S_0 = the initial surfactant concentration;
ρ_w = the radical flux in the aqueous phase;
μ = the volumetric growth rate of the polymer particles.

The Smith–Ewart treatment of Interval II is based on a particle balance over the reactor:

$$\frac{dN_i}{dt} = \underbrace{[N_{i-1} - N_i]\left(\frac{\rho_a}{N}\right)}_{\text{(radical entry)}} + \underbrace{\frac{k_0 a_p}{V}[(i + 1)N_{i+1} - iN_i]}_{\text{(radical desorption)}}$$

$$+ \underbrace{\frac{k_p}{VN_a}[(i + 2)(i + 1)N_{i+2} - i(i - 1)N_i]}_{\text{(mutual termination)}} \qquad (4.2)$$

Here N_i is the concentration of particles containing i radicals, ρ_a is the flux of radicals entering the particles, N is the total number concentration of particles,

a_p and V are respectively, the area and volume of a particle, N_a is Avogadro's number, and k_0 is the rate constant for radical absorption.

When the steady-state approximation is made, the rate of polymerization becomes

$$R_p = k_p M_p \sum_{i=0}^{\infty} i \frac{N_i}{N_a} = k_p M_p \bar{i} \frac{N}{N_a} \tag{4.3}$$

where $\bar{i}$ is the average number of radicals per particle and M_p is the monomer concentration in the polymer particles.

Instead of solving the particle balance [Equation (4.2)] directly, Smith and Ewart treated three limiting cases:

Case I: $\bar{i} \blacktriangleleft 1$
Case II: $\bar{i} = 0.5$
Case III: $\bar{i} \blacktriangleright 1$

Case II corresponds to the situation in which a radical propagates in a particle until another radical enters, at which time instantaneous termination takes place and the particle lies dormant until a third radical enters the particle and reinitiates polymerization. Thus, each particle contains a radical half the time, so that the average number of radicals per particle is 0.5. Data for styrene and other monomers have been found to follow this rate law. Case I kinetics are found for highly water-soluble monomers such as vinyl acetate. Case III kinetics occur when the particles are very large (the system approaches suspension kinetics) or when the number of radicals per particle is high due to reduced termination due to the gel effect. A great deal of data which matches the Smith–Ewart model has been reported over the years, primarily for highly water-insoluble monomers like styrene. Results for these and other systems which deviate the Smith–Ewart model have been tabulated by Min and Ray [5].

Using the steady-state approximation, Stockmeyer [15] has solved the particle balance [Eq. (4.2)] analytically to give the number of radicals per particle. His work has been modified by O'Toole [16].

For Case II kinetics, the instantaneous degree of polymerization can be written (as before) as the rate of propagation divided by the rate of termination:

$$\bar{x} = \frac{k_p}{\rho_w/N} = \frac{k_p M_p N}{\rho_w} \tag{4.4}$$

Referring back to Eqs. (1.46) and (1.48), it may be seen that in a bulk (or solution, or suspension) polymerization, if the initiator level is increased to increase the rate of polymerization, the NACL will fall. On the other hand, inspection of Eqs. (4.3) and (4.4) will indicate that the rate of polymerization can be increased by increasing the surfactant level (thus creating more particles) while increasing the NACL simultaneously. It is the ability to produce very high-molecular-weight material (above 10^6) at high reaction rates that makes emulsion polymerization the process of choice for many applications.

4.2.3 Reactor Modeling, Operation, and Dynamics

Batch emulsion polymerizations are typically carried out in large (1000–100,000 liter) agitated vessels. These reactors may be operated isothermally using a cooling jacket to remove the heat of polymerization, or they may be operated nonisothermally with fixed cooling. If the kettle is heat transfer limited, the initiator may be added over the course of the polymerization to keep the heat evolution within the capacity of the cooling system. In emulsion copolymerization, it may be necessary to add the more reactive (or more water-soluble) monomer in a semibatch mode to obtain the correct copolymer composition distribution. The polymerization may be stopped well short of completion to reduce chain branching which is especially prevalent for some monomers when polymerized in an emulsion system. In this case, the residual monomer is recovered from the latex and recycled. On the other hand, the charge may be dropped from the kettle before it reaches complete conversion and allowed to react to complete conversion in a storage tank. This is done to reduce kettle time and may be practical because, near the end of the polymerization, the heat evolution (and rate of polymerization) are quite small. Multiple surfactants and multiple initiators may be used to obtain desired colloidal or molecular properties. Monomer conversion may be tracked by gravimetric measurements [17]. A comprehensive simulation model of batch emulsion polymerization has been developed by Min and Ray [5].

High-volume products such as styrene-butadiene rubber (SBR) often are produced by continuous emulsion polymerization. This is most often done in a train of 5–15 CSTRs in series. Continuous emulsion polymerization has some unique features which make understanding and optimizing the process particularly worthwhile. The reaction history of a polymer is written into its molecular weight distribution. Sustained oscillations (limit cycles) in conversion, particle number, and free emulsifier concentration have been reported [18–21] under isothermal conditions in continuous emulsion polymerization systems. This limit cycle behavior leaves its mark on the product in the form of disturbances in the molecular weight distribution which cannot be blended away. In addition, multiple steady states have been observed [21,22] compounding the problems of reactor control and product quality.

The earliest model for continuous emulsion polymerization was that of Gershberg and Longfield [18] for styrene. Their model is an adaptation of the Smith–Ewart model to a series of continuous stirred tank reactors. By combining balances over a number of particles and total emulsifier in each reactor, with the Smith–Ewart equation for the rate of polymerization and the residence time distribution for the CSTR, then making the QSSA for all species, they were able to develop equations for the number of particles and rate of polymerization in each stage. They report data for a series of three CSTRs in which conversion and particle size were measured. Their data for styrene show a transient maximum in conversion followed by a rapid decline to the steady state. In addition, with some combinations of operating variables, they report steady oscillations in conversion of amplitude of 1–4% and period of 3–5 residence times. Since theirs is only a steady-state model, they cannot, of course, predict these transients. They do, however, predict steady-state conversion, rate of polymerization, and particle number with reasonable accuracy. More complex models of continuous emulsion polymerization have been developed. The best of these appears to be that of Kiparissides [23].

As noted above, continuous emulsion polymerization reactors can exhibit unusual and undesirable dynamics. These include a multiple steady state of much the same sort found in CSTR free radical solution polymerization, and damped, or continuous, oscillations.

Figure 4.2 is a plot of the steady-state monomer conversion as a function of reactor residence time for methyl methacrylate emulsion polymerization [24]. A region of multiplicity is indicated by the fact that the upper and lower branches of the curve overlap between residence times of 30 and 50 min. The dotted line is an estimate of the shape of the unstable middle branch which is experimentally unobservable. The dashed lines indicate experimental instances of ignition and extinction. At 50 min residence time, the system has been observed to move from the lower steady state of 54% conversion to the upper steady state at approximately 80% with no discernible change in operating conditions (ignition). Extinction has been observed when the residence time is changed from 30 min to 20 min on the upper branch resulting in a drop in conversion from the upper to the lower steady-state values. The phenomenon of multiple steady states arises in emulsion polymerization for much the same reason it appears in solution polymerization: the autocatalytic nature of the polymerization (due to the gel effect) combined with the mass balance results in the possibility of steady state multiplicity.

Figure 4.3 shows evidence of a sustained oscillation (limit cycle) during emulsion polymerization of methyl methacrylate in a single CSTR. Comparison of the monomer conversion and surface tension data graphically illustrates the mechanism of oscillation. It will be noted that the surface tension oscillates with the same period as the conversion (six to seven residence times). This is in agreement with the classical micellar initiation mechanism. Beginning at a time of about 300 min, the conversion rises rapidly as new particles form and old particles grow. As the particle surface area increases, additional surfactant is adsorbed on the

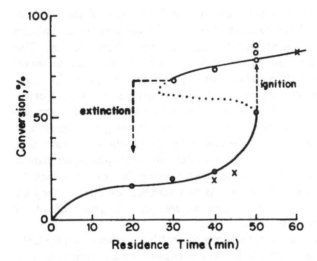

Figure 4.2 Steady-state multiplicity in the continuous emulsion polymerization of methyl methacrylate. (From Ref. 24 with permission.)

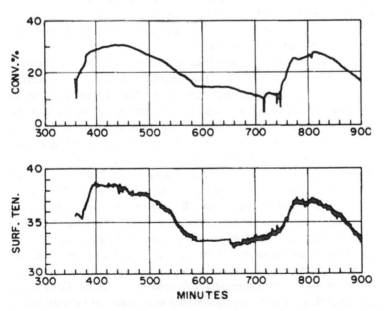

Figure 4.3 Limit cycle oscillation in the continuous emulsion polymerization of methyl methacrylate. (From Ref. 24 with permission.)

particles. Meanwhile, micelles dissociate to keep the aqueous phase saturated. Once all of the micelles have dissociated, it is no longer possible to maintain the aqueous phase at saturation, and the surface tension begins to rise. This is observed at about 320 min. At the point at which micelles are no longer present, micellar initiation stops and the rate of polymerization slows. Eventually, because particles are washing out while no new particles are being formed, the conversion begins to fall. Because the total particle surface area is decreasing at this point and because surfactant is continually being introduced with the feed, the surface tension falls as the aqueous phase reapproaches saturation. As the aqueous phase becomes saturated, micellar initiation begins again. Saturation of the aqueous phase may be observed by noting the point at which the surface tension reaches its CMC value. As new micelles are formed, they adsorb free radicals, become polymer particles, and begin to grow and adsorb surfactant. The cycle then repeats.

More recent results [25] show that whereas the instability arises above the CMC (and is promoted by large values of initiator concentration and residence time, and low surfactant concentration), it is the on/off nature of the micellar nucleation mechanism which governs the nature of oscillations in monomer conversion. The surface tension oscillation leads the conversion oscillation by approximately one residence time. This is consistent with the above explanation because changes in surfactant concentration are quite rapid, whereas changes in the number of particles and rate of reaction require a finite growth time to appear as changes in the monomer conversion. Damped oscillations upon start-up have been noted for a large number of monomer systems.

Multiplicity is undesirable in a commercial reactor because ignition or extinction may occur unexpectedly. This may lead to loss of temperature control in the case of ignition, or loss of reactor productivity in the case of extinction. Oscillations (both damped and sustained) are undesirable because the product is not of a consistent quality and because oscillations in free surfactant concentration may induce coagulation and reactor fouling. Several methods of eliminating oscillations have been suggested. Geene et al. [20] have used a plug flow reactor upstream of a CSTR train. All polymer particles are nucleated in the PFR. Because PFR kinetics are essentially those of a batch reactor, no oscillations occur. The CSTRs, then, are used to grow the existing particles. By segregating particle nucleation from particle growth, oscillations are eliminated. Another approach has been taken by Penlidis [26] and others. This involves using a small CSTR as a seeder reactor. All polymer particles are formed in the seeder. A portion of the monomer and water is bypassed around the seeder in such a way as to dilute out any remaining micelles in the reactor immediately following the seeder. Once again, nucleation and growth have been segregated and oscillations are eliminated. A good review of emulsion polymerization reactor design, operation, and control is given by Hamielec and MacGregor [27].

4.3   Coordination Polymerization

Highly linear stereospecific polyolefins are produced by a class of heterogeneous polymerization processes known as *coordination* or stereospecific polymerization. This process accounts for a very large portion of all synthetic polymers produced. Because Ziegler–Natta catalysts are most often used, the processes are often loosely referred to as Ziegler–Natta polymerizations. Two distinct processes are commercially important: slurry polymerization and gas phase polymerization. In both the slurry and gas phase processes, the monomers are gaseous olefins, especially ethylene and propylene. For the production of polypropylene, coordination polymerization is used out of necessity because propylene will not polymerize by radical or ionic mechanisms. The coordination mechanism produces highly linear polymer with monomers such as ethylene which tend to produce highly branched material when polymerized by a free radical mechanism. The linearity is due to the low temperature of polymerization and the relative absence of transfer to the polymer backbone. The primary advantage of coordination polymerization lies in its ability to produce stereoregular polymer from alpha-olefins. Polypropylene may be produced in either an isotactic or syndiotactic form, depending on the choice of catalyst, but higher alpha-olefins produce only the isotactic form. With proper catalyst design, dienes may be polymerized to polymers which are almost exclusively *cis*-1,4, *trans*-1,4, isotactic 1,2, or syndiotactic 1,2.

4.3.1   Process

In both the slurry and gas phase processes, the same types of catalysts (Ziegler–Natta catalysts) are used. These are complexes of metal alkyls of groups I–III (such as aluminum alkyl) with halides of group IV–VIII metals (such as titanium chloride). In both processes, the catalyst forms a solid, particulate phase. The polymerization takes place on the catalyst surface.

In the slurry process, the catalyst is slurried in an organic solvent. Gaseous monomer under pressure is introduced into the reactor and dissolved in the liquid phase. Polymerization takes place in the liquid phase at the surface of the catalyst particle. The polymer (insoluble in the liquid phase) forms a slurry and is recovered by filtration at the end of the polymerization.

In the gas phase process, the organic solvent is eliminated, and the catalyst particles are immersed in a fluidized or stirred bed of gaseous monomer. Polymerization takes place in the gas phase, at the particle surface. The polymer particles thus formed are then recovered from the gas phase. In the slurry process, removal of the heat of reaction is done by cooling and recycling the gas phase. Because heat exchange with a gas is not particularly efficient, temperature control is often a problem with gas phase systems.

4.3.2 Kinetics

The kinetics of coordination polymerization are similar to those of ionic polymerization with the exception that they take place on the surface of the catalyst particle. A general kinetic scheme may be written as follows [28]:

$$AR \overset{K_1}{\Leftrightarrow} A^*R \Biggr\}$$ (4.5)

$$\phantom{AR \overset{K_1}{\Leftrightarrow} A^*R} \text{Adsorption}$$

$$M \overset{K_2}{\Leftrightarrow} M^* \Biggr\}$$ (4.6)

$$M^* + A^*R \overset{k_i}{\Rightarrow} A^*P_1R \qquad \text{Initiation} \tag{4.7}$$

$$M^* + A^*P_nR \overset{k_p}{\Rightarrow} A^*P_{n+1}R \qquad \text{Propagation} \tag{4.8}$$

$$M^* + A^*P_nR \overset{k_t}{\Rightarrow} A^*M + M_nR \qquad \text{Termination} \tag{4.9}$$

$$M^*A^*P_nR \overset{k_{fm}}{\Rightarrow} A^*P_1R + M_n \qquad \text{Chain Transfer to Monomer} \tag{4.10}$$

Here AR is the metal alkyl, (*) indicates that the species is adsorbed on the transition metal, and P_n and M_n are interpreted as previously. The metal alkyl forms a complex with the surface of the transition metal. The monomer adsorbs onto the surface, and initiation takes place as the monomer is inserted into the complex. Propagation takes place as more monomer is inserted into the chain. Note that the chain remains attached to the catalyst surface. Termination may take place with the formation of an inactive site. Chain transfer to monomer (or other species) controls the chain length.

If adsorption is assumed to follow a Langmuir isotherm and the QSSA is made for growing chains, the rate of polymerization may be written as

$$R_p = \frac{(k_p + k_{fm})k_i}{k_t} \frac{K_1 K_2 (AR)M}{[1 + K_1(AR) + K_2 M]^2} \tag{4.11}$$

where the species symbols are now interpreted as concentrations. The instantaneous degree of polymerization may be written as previously, as the ratio of the rate of propagation to the sum of the rates of termination and chain transfer. With the simple kinetic model above, this results in

$$\bar{x} = \frac{k_p}{(k_t + k_{fm})} \tag{4.12}$$

The observed kinetics are often much more complex than the above. Mass transfer effects may be significant. In addition, it is difficult to determine the number of active sites, and thus K_1 and K_2. Reactivity is often expressed as grams of polymer per gram of catalyst. Because the catalyst remains with the polymer, one of the major goals of catalyst design is to improve this ratio.

Coordination polymerization, like ionic polymerization, is extremely sensitive to inhibition by water and oxygen. Hydrogen may be used as a chain transfer agent to control molecular weight. Programmed adjustments of the monomer and hydrogen partial pressures may be used to tailor polymer properties. A more extensive discussion of coordination polymerization is available in Odian [29].

References

1. E. Trommsdorff and E. Schildknecht, "Polymerizations in Suspension," in *Polymer Processes*, Interscience, New York, 1956.
2. W. D. Harkins, "A General Theory of the Reaction Loci in Emulsion Polymerization," *J. Chem. Phys.*, *13*, 381 (1945).
3. W. D. Harkins, "A General Theory of the Reaction Loci in Emulsion Polymerization. II," *J. Chem. Phys.*, *14*, 47 (1946).
4. W. D. Harkins, "A General Theory of the Mechanism of Emulsion Polymerization," *J. Amer. Chem. Soc.*, *69*, 1428 (1947).
5. K. W. Min and W. H. Ray, "On the Mathematical Modeling of Emulsion Polymerization Reactors," *J. Macro. Sci.* (*Revs.*), *11*, 177 (1974).
6. R. M. Fitch and C. H. Tsai, "Particle Formation in Polymer Colloids, III: Prediction of the Number of Particles by a Homogeneous Nucleation Theory," in *Polymer Colloids* (R. M. Fitch, ed.), Plenum, New York, 1971.
7. J. Ugelstad, P. C. Mork, and J. O. Aasen, "Kinetics of Emulsion Polymerization," *J. Polym. Sci.*, *A-1*, *5*, 2281–2288 (1967).
8. B. W. Brooks, "Particle Nucleation Rates in Continuous Emulsion Polymerization Reactors," *Br. Polym. J.*, *5*, 199 (1973).
9. D. H. Napper, A. Netschey, and A. E. Alexander, "Effects of Nonionic Surfactants on Seeded Polymerizations," *J. Polym. Sci.* (*Chem.*), *9*, 81 (1971).
10. H. J. Van den Hul and J. W. Vanderhoff, "The Characterization of Latex Particle Surfaces by Ion Exchange and Conductometric Titration," *J. Electroanal. Chem.*, *37*, 161 (1972).
11. B. R. Vijayendran, "Monomer Polarity and Surfactant Adsorption," *Appl. Polym. Sci.*, *23*, 733 (1979).
12. W. V. Smith, "The Kinetics of Styrene Emulsion Polymerization," *J. Amer. Chem. Soc.*, *70*, 3695 (1948).
13. W. V. Smith and R. H. Ewart, "Kinetics of Emulsion Polymerization," *Chem. Phys.*, *16*, 592–599 (1948).
14. W. V. Smith, "Chain Initiation in Styrene Emulsion Polymerization," *J. Amer. Chem. Soc.*, *71*, 4077 (1949).
15. W. H. Stockmayer, "Note on the Kinetics of Emulsion Polymerization," *J. Polym. Sci.*, *24*, 314 (1957).
16. J. T. O'Toole, "Kinetics of Emulsion Polymerization," *J. Appl. Polym. Sci.*, *9*, 1291 (1965).

17. F. J. Schork and W. H. Ray, "On-Line Monitoring of Emulsion Polymerization Reactor Dynamics," in *Emulsion Polymers and Emulsion Polymerization*, (D. R. Bassett and A. E. Hamielec, ed.), American Chemical Society, Washington, D.C., 1981.
18. D. B. Gershberg and J. E. Longfield, "Kinetics of Continuous Emulsion Polymerization," Paper 10, 45th AlChE Meeting, New York, 1961.
19. M. Nomura and M. Harada, "On the Optimal Reactor Type and Operations for Continuous Emulsion Polymerization," in *Emulsion Polymers and Emulsion Polymerization* (D. R. Bassett and A. E. Hamielec, eds.), American Chemical Society, Washington, D.C., 1981.
20. R. K. Greene, R. A. Gonzalez, and G. W. Poehlein, "Continuous Emulsion Polymerization — Steady State and Transient Experiments with Vinyl Acetate and Methyl Methacrylatem" in *Emulsion Polymerization* (I. Piirma and J. Gardon, eds.), American Chemical Society, Washington, D.C., 1976.
22. G. Ley and H. Gerrens, "Mehrfache stationare Zustande und periodische Teilchenbildung bei der Emulsionpolymerisation von Styrol im kontinuierlichen Ruhrkesselreactor," *Makromol. Chem.*, *175*, 563 (1974).
23. C. Kiparissides, J. F. MacGregor, and A. E. Hamielec, "Continuous Emulsion Polymerization of Vinyl Acetate Part: II," *Can. J. Chem. Eng.*, *58*, 56 (1980).
24. F. J. Schork and W. H. Ray, "The Dynamics of Continuous Emulsion Polymerization of Methylmethacrylate," *J. Appl. Polym. Sci.*, *34*, 1259–1276 (1987).
25. J. B. Rawlings, "Simulation and Stability of Continuous Emulsion Polymerization Reactors," Ph.D. Thesis, University of Wisconsin, 1985.
26. J. F. Penlidis, J. F. MacGregor and A. E. Hamielec, "Continuous Emulsion Polymerization Reactor Control," *Proceedings of the 1985 American Control Conference*.
27. A. E. Hamielec and J. F. MacGregor, "Latex Reactor Principles: Design, Operation, and Control," in *Emulsion Polymerization*, (I. Piirma, ed.), Academic Press, New York, 1982.
28. F. W. Billmeyer, Jr., *Textbook of Polymer Science*, 3rd ed., Wiley, New York, 1984.
29. G. Odian, *Principles of Polymerization*, 2nd ed., Wiley, New York, 1981.

5

Framework of the Control Problem

The production objectives in polymerization are to produce a polymer having certain properties for a specific application. The nature of the application, along with process economics, determines what raw materials [monomer(s), catalysts] are appropriate, but the polymer with desired properties will not result unless reaction conditions are tightly controlled. The development of control strategies for polymerization reactors requires an appreciation of what the important properties are and how they relate to variables within the reactor and, furthermore, what inputs are available with which the reactor variables may be regulated at desired levels. In this chapter, we consider polymerization reactions and develop a fundamental basis for controlling polymerization reactors.

5.1 Introduction

One of the first considerations in establishing a strategy for controlling polymerization reactors is to categorize all system inputs and outputs into those which are to be controlled, those which may be adjusted to achieve this control, and those which are beyond the control of the designer. The cause-and-effect relationships in a polymerization reactor are depicted in Fig. 5.1.

The *system outputs* may be divided into three catagories: *end-use properties, controlled variables affecting product quality,* and *controlled variables specifying operating conditions.* In many instances, the polymer prepared in a reactor

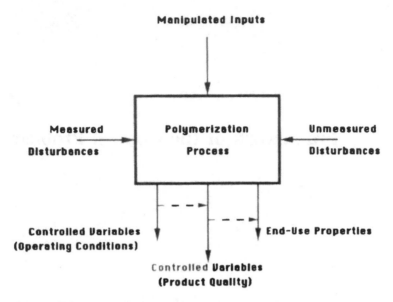

Figure 5.1 Polymerization process variables.

must be further processed to produce a desired product. So that the final product may meet the required specifications for a given application, the polymer must have certain *end-use properties*. End-use properties determine the suitability of a polymer for a specific application. These properties may be well-defined, e.g., tensile strength, or they may be empirical measures of suitability in a given application. End-use properties such as solubility, bulk density, extrudability, etc., determine the salability of the polymer. Polymer appearance factors such as color, refractive index, particle size, and shape, are also important in some cases. In most cases, the end-use properties are not measurable on-line. Other outputs must be controlled so as to produce a polymer having the desired end-use properties.

From a control standpoint, the most important variables are those which ultimately affect the end-use properties of a polymer. We refer to these as *controlled variables affecting product quality*. The most important of these are MW, MWD, monomer conversion, copolymer composition distribution (CCD), copolymer sequence distribution, and degree of branching. Most of these variables are also not measurable on-line. The approach is to closed-loop control those variables which are measurable, to estimate those which are not and control based on the estimates, and to open-loop control those which cannot be estimated. By closed-loop control is meant the adjustment of some manipulated variable(s) in response to a deviation of the associated control variable from its desired value.

The purpose of closed-loop control is to bring the controlled variable to its desired value and maintain it at that point. By open-loop control is meant the control of all identifiable inputs to maintain an unmeasured output at a constant value.

The next category of process outputs are called *controlled variables specifying operating conditions*. Some examples are temperatures, pressures, and flow rates, associated with the process. These variables are most often measurable and are closed-loop controlled.

The inputs to the polymerization system can be categorized as *manipulated variables* or *disturbance variables*. The *manipulated variables* are those which are adjusted, either automatically or manually, to maintain the controlled variables at their desired values. Common manipulated variables in polymerization processes include coolant or heating medium flow rates, gas or liquid flow rates for pressure control, feed rate of monomer, solvent, or initiator, and agitator speed.

The *disturbance variables* are, by definition, those over which the control engineer has no control. Disturbances may be stochastic (random) or deterministic. Stochastic disturbances arise from the natural variability of the process. Examples are short-term variations in flow rates caused by mechanical inaccuracies or variability due to day-to-day variations in feedstock quality. Deterministic disturbances arise from known causes and they usually occur at longer intervals. Examples are lot-to-lot variations in feedstock quality and changes in production rates mandated by the operation of some upstream or downstream process. Although the cause of such disturbances is usually known, the disturbances themselves cannot be eliminated because of constraints external to the system. Some disturbances, stochastic and deterministic, may be measurable, but, by definition, they cannot be eliminated. However, the *effect* of such disturbances on the final product can be eliminated by compensating for them by adjusting the manipulated variables. This is the function of regulatory control.

The process variables do not always fall into such neat categories. For instance, temperature may be manipulated to adjust MW. In this case temperature is the manipulated variable for an MW control loop, but may at the same time be the controlled variable for a temperature control loop which uses the flow rate of a coolant as a manipulated variable. In this case, the value of the manipulated variable for the MW control loop (temperature) is the desired value for the temperature control loop. This, of course, is the notion of cascade control.

Not all process variables are measurable on-line. Most end-use properties are not. Some controlled variables affecting product quality are measurable (or they can be estimated); many are not. Even when the technology to measure these variables on-line exists, the cost of such sensors may be prohibitive; on-line gel permeation chromatography for the determination of MWD is an example. Almost all controlled variables specifying operating conditions are measured. Most manipulated variables are known because they are either measured or set by a control system (or both). As discussed above, disturbances may be measured or unmeasured.

One final point about closed-loop process control. Economic considerations dictate that to derive optimum benefits, processes must invariably be operated in the vicinity of constraints. A good control system must drive the process toward these constraints without actually violating them. For example, a distillation control system can maximize the quantity of feed processed by manipulating the reflux flow and reboiler heat. However, transient disturbances could cause flooding. The control system then would be required to operate the column in the vicinity of the flooding constraint so as to maximize productivity under normal operation, but should flooding become imminent, it would be allowed to give up the productivity objective and shift priorities so that stable operation is maintained. The manipulated variables may also have their own constraints. For example, the reflux flow must not be allowed to fall beyond a certain minimum or exceed a certain maximum limit; the maximum rate of change of a manipulated variable permitted within one sampling period may be limited to a certain value, etc. Likewise, in a polymerization reactor, the initiator feed rate may be manipulated to control monomer conversion or MW; however, at times when the heat of polymerization exceeds the heat transfer capacity of the kettle, the initiator feed rate must be constrained in the interest of thermal stability. In some instances, there may be constraints on the controlled variables as well. Identification of constraints for optimized operation is an important consideration in control systems design. Operation in the vicinity of constraints poses problems because the process behavior in this region becomes increasingly nonlinear.

Depending on the quantity of polymer to be produced, polymerization may be carried out either in a batch reactor or in a CSTR (continuous stirred tank reactor). Operational strategies for a batch reactor would differ from those for a CSTR even if the desired product is the same. Based on our understanding of polymerization principles and in the light of our discussion of system inputs and outputs, a preliminary description of batch and continuous polymerization reactor control strategies may be developed as we will see in the following paragraphs. The goal of the control system for either type of reactor would be to maintain certain outputs within specifications by adjusting suitable inputs, subject to constraints in the presence of disturbances.

5.2 Control of Batch Polymerizations

Batch or semibatch polymerization control systems commonly include preprogrammed recipe addition and start-up and shut-down procedures. Batch polymerizations are commonly carried out in a vessel under temperature control, either isothermally or following some predetermined temperature profile. Cooling to remove the rather large heat of polymerization is normally provided by a coolant flowing in the jacket. If a very broad MWD is desired (as in adhesives), the reactor may be run adiabatically, although care must be taken to avoid runaway. If any of the monomers or solvents have high vapor pressures, reactor pressure may be controlled as well. This is often done with a nitrogen blanket in the head space of the reactor. This has the added benefit of eliminating oxygen which may retard initiation.

All of the initiator or catalyst may be added at the beginning of the reaction or some may be reserved for addition over the course of the polymerization. If some of the initiator is added over the course of polymerization, the addition may be prescribed according to a predetermined trajectory or it may be adjusted to maintain a constant rate of polymerization over most of the course of polymerization. Alternatively, if an on-line MWD measurement is available, initiator addition may be used to regulate MW, although this is rarely done.

In copolymerization, the more reactive monomer may be added to the reactor over time to produce a more uniform copolymer composition distribution. If copolymer composition is measured or estimated on-line, the reactive monomer can be added in a closed-loop fashion [5].

In emulsion polymerization, a surfactant may be added over time to control the formation of new particles and, hence, the particle size distribution (PSD) [6].

The determination of trajectories for the addition of monomers or initiators, and/or temperature in batch or semibatch polymerizations are almost always done off-line. These trajectories may be the result of operating experience or they may be developed by calculating optimal trajectories to achieve certain goals (reduced kettle time, desired MWD or CCD) subject to constraints on heat transfer capacity, etc. [7]. The rigorous calculation of optimal trajectories requires a reasonably accurate model of the polymerization process. On-line optimization of operating trajectories is often not done due to the lack of detailed mathematical models and the computational time necessary to develop an optimal trajectory.

5.3 Control of Continuous Polymerizations

In continuous polymerizations, temperature and pressure are controlled in much the same way as in batch systems. Obviously, temperature trajectories are not employed in continuous reactors. Instead, the various vessels in a series of polymerization reactors may operate at different temperatures. The polymerizing mixture, then, will "see" different temperatures as it passes from one vessel to the next. Likewise, monomer trajectories are replaced, in continuous systems, with intermediate injection of a more reactive monomer between polymerization vessels. This strategy can be exploited to adjust the copolymer composition distribution.

Monomer conversion can be adjusted by manipulating the feed rate of the initiator or catalyst. If on-line MWD is available, the initiator flow rate or reactor temperature can be used to adjust MW [8].

In emulsion polymerization, initiator feed rate can be used to control monomer conversion, whereas bypassing part of the water and monomer around the first reactor in a train can be used to control PSD [9,10]. Direct control of surfactant feed rate, based on surface tension measurements, also can be used.

Optimal policies for changing product grades without shutting down the reactor train are desirable, but what little is done in this area is based on operating experience rather than on optimization approaches.

5.4 Control Topics

We end this chapter with a synopsis of topics to be discussed in the remainder of the text. We begin in the next chapter with a discussion of the measurement and estimation of process variables in polymerization. This will be followed by a chapter on process modeling and identification. Model-based control has become increasingly important and the material in this chapter will be useful in the development of model-based control systems. In our previous discussion, we have seen that there are several inputs and outputs in a polymerization system. Thus, a rigorous system for controlling polymerization reactors would be multivariable in nature. However, in certain situations, the polymerization reactors can and are operated with single-input single-output (SISO) control strategies. The goal of the following chapter then is to present state-of-the-art SISO control algorithms and their applications to polymerization reactor control. This will be followed by a chapter devoted to multivariable control techniques (multiple-input multiple-output, or MIMO). To derive maximum benefits, a polymerization reactor may have to be operated with a multivariable control system having "constraint handling" capabilities. In the chapter, we present a systematic methodology and techniques for multivariable control and their applications to polymerization systems. In the early part of this text we have seen that polymerization reaction systems are described by nonlinear differential equations. It is anticipated that nonlinear control methodologies will, therefore, yield superior control. In the next chapter we cover the recent developments in nonlinear control that appear to hold some promise for polymerization reactor systems.

In a large number of applications, the polymer produced in a reactor is not the final product. The final chapter of the text introduces the reader to polymer processing operations. The material in this chapter should be useful because the requirements of the polymer processing system may dictate operational changes in the polymerization reactor control system.

References

1. W. H. Ray, "Polymerization Reactor Control," Proceedings of the *1984 American Control Conference*, Boston, June, 1984.

2. H. Amrehn, "Computer Control in the Polymerization Industry," *Automatica*, *13*, 533–545 (1977).
3. K. Hoogendorn and R. Shaw, "Control of Polymerization Processes," in Fourth IFAC Conference on Instrumentation and Automatic Control in the Paper, Rubber, Plastics and Polymerization Industries, Ghent, Belgium, June, 1980.
4. J. F. MacGregor, A. Penlidis, and A. E. Hamielec, "Control of Polymerization Reactors: A Review," *Polym. Process Eng.*, *2*, 179–206 (1984).
5. A. Guyot, J. Guillot, C. Pichot, and L. R. Guerrero, "New Design for Producing Constant-Composition Copolymers in Emulsion Polymerization: Comparison with Other Processes," in *Emulsion Polymers and Emulsion Polymerization* (D. R. Bassett and A. E. Hamielec, eds.), American Chemical Society, Washington, D.C., 1981.
6. D. L. Gordon and K. R. Weidner, "Control of Particle Sized Distribution Through Emulsifier Metering Based on Rate of Conversion," in *Emulsion Polymers and Emulsion Polymerization* (D. R. Bassett and A. E. Hamielec, eds.), American Chemical Society, Washington, D.C., 1981.
7. A. E. Hamielec and J. F. MacGregor, "Thermal and Chemical-Initiated Copolymerization of Styrene/Acrylic Acid at High Temperatures and Conversions in a Continuous Stirred Tank Reactor," in *Proceedings of the International Berlin Workshop on Polymer Reaction Engineering*, Berlin, October, 1983.
8. A. K. Adebekun and F. J. Schork, "Continuous Solution Polymerization Reactor Control. 1. Nonlinear Reference Control of Methylmethacrylate Polymerization," *I & EC Research*, *28*, 1308–1324 (1989).
9. M. J. Pollock, J. F. MacGregor, and A. E. Hamielec, in *Computer Applications in Applied Polymer Science* (T. Provder, ed.), ACS Symposium Series Volume 197, American Chemical Society, Washington, D.C., 1981, pp. 209–220.
10. K. O. Temeng and F. J. Schork, "Closed-Loop Control of a Seeded Continuous Emulsion Polymerization Reactor System," *Chem. Eng. Commun.*, *85*, 193–219 (1989).

6

Measurement and Estimation of Process Variables

In this chapter, the impact of measurement techniques and capabilities on the design of control systems for polymerization will be considered. In many cases, the capability of controlling polymerizations is severely limited by the state of the art in measurement instrumentation. In other cases, the dynamic response of the instruments dictate the design strategy for the process. These considerations will be discussed below for those variables most generally measured in polymerization processes. A detailed discussion of measurement systems (except composition measurement) is given by Doebelin [1].

6.1 Measurement of Temperature

In nearly every chemical reaction, the temperature of the reacting mass is of utmost importance. Polymerization is no different; control of reaction temperature or, in some cases, the temperature–time profile is essential. Most polymerizations are exothermic and require maintenance of reaction temperature for reasons of process safety, if for no other. More typically, however, the reaction temperature influences final polymer properties and, therefore, must be regulated.

In general, temperature measurements for polymerization reactions are made with the same instrumentation that is generally used in the chemical industry. Three classes of instruments are normally used: thermocouples, resistance thermometers, and filled-bulb thermometers. All of these devices display dynamic

Table 6.1 Dynamics of Temperature Measurement

Device	Time constant (sec)
Thermocouple	30–60
RTD	<5
Filled Bulb	60–120

responses which can be characterized by a first-order lag model without dead-time. The time constants for these devices are typically in the range shown in Table 6.1.

6.2 Measurement of Pressure

Measurement of pressure is important in many polymerization processes. Typically, in exothermic closed reactor systems, pressure is monitored to ensure that the reaction is being safely conducted. Also, in some systems, molecular configuration and reaction rate are pressure dependent. In some special cases where reaction takes place in a melt extruder, measurement of pressure at the discharge can be related to viscosity or molecular weight. Each particular polymerization has different demands on the pressure-measuring instruments in terms of required precision and noise level. In general, pressure is measured by three basic techniques: (1) liquid columns (characterized by the water or mercury manometer), (2) mechanical sensing elements (diaphragm, bellows, or Bourdon tube), and (3) electrical or electronic methods (strain gauges, thermal-conductivity gauges, or ionization gauges). Each of these methods is used for different pressure ranges and situations, but, in general, these instruments are very fast in response. In fact, most of the pressure sensor/transmitter systems which are used in the chemical industry have response times of a second or less. Compared to the process lags encountered in most polymerizations, the dynamics of pressure measurement can generally be considered insignificant.

6.3 Measurement of Flow and Weight

A key consideration in conducting polymerization is the maintenance of the proper recipe or ingredient mix. In most polymerizations and all copolymerizations, more than a single ingredient is required in the reacting mass. For example, in some emulsion polymerizations, as many as 20 different components must be fed into the reaction at controlled concentrations.

In batch systems, the concern is to charge the reactor as quickly as possible with the correct recipe. Charging of the reactor is typically done in two ways:

1. Monitoring the reaction mass with a scale or weigh cell and adding the components one at a time. When the weight of the desired component is reached, a digital signal is sent to the flow valve. This type of operation lends itself to control with a programmable controller.
2. Monitoring of the flow rate of each ingredient into the batch and integration over time to give totalized flow. Again, as the desired total mass of each component is reached, the flow valve is closed.

For continuous systems, the flow rates of ingredients are controlled. Achievement of the desired composition requires ratio controllers in which all flow rates are tied to a single key component for feedforward control. In addition, flow-rate setpoints may be remotely set in a cascade fashion from a master composition loop.

In either case of weight or flow-rate measurement, the major problems are not due to the dynamics of the instruments because they are extremely fast compared to process dynamics. More typically, however, problems arise because of other factors including:

1. The nature of the generated signal. Often the instrument produces a signal with a large amount of high-frequency noise, which must be filtered out prior to use in the control system. Although dynamically fast, the flow-rate measurements are inherently noisy and care must be taken in the proper pretreatment of the signal.
2. The nature of the monitored species. In some cases it is not possible to measure the flow rate with standard differential pressure (DP) measurements because the material being monitored plugs the orifice or is too viscous. Novel measurement methods, such as ultrasonic, turbine, or target meters, must then be used.

6.4 Measurement of Composition or Conversion

Up to this point, all of the measurements which have been discussed are generally very fast compared to the dominant process time constants. In the case of measurement of composition (or conversion of monomer to polymer), however, this generalization may no longer hold. There are basically two reasons why composition measurement is needed in polymerization systems:

1. to ensure that the proper combination of ingredients is available in the reaction mass at the beginning of or in the course of the polymerization;
2. to determine the extent of reaction (or monomer conversion) which has taken place in the reacting mass in real time to facilitate reaction control.

In the first case, the problem is generally to measure the concentration of one or more chemical species in the reaction mixture. Most often, this problem is quite readily solved by the development of a suitable measurement technique based on a specific property of the component to be measured. The most frequently used techniques include spectroscopy, colorimetry, light scattering, titration, chromatography, refractive index, density, thermal conductivity, and vapor pressure [2–4].

Generally, the problem is not development of a suitable analysis, but application of that analysis to on-line measurement. This problem is most clearly seen in the case of batch polymerization. Suppose that the composition of one critical component in the batch makeup must be known precisely before the addition of a catalyst or initiator which begins the polymerization. Without an on-line instrument to measure that composition, a sample must be taken, sent to the lab, analyzed, and reported, before any further action can be taken. This analysis time represents a true loss of utility of the polymerization equipment. On the other hand, in the case of a continuous polymerization system, in which composition control is essential, the lack of an on-line instrument creates the classical deadtime problem: The operator must wait until analyses are available and then make cautious adjustments because the process state could have changed between the time the sample was taken and the time the analysis becomes available. Composition measurement is often the major source of deadtime in composition control loops.

In the area of composition measurement, we have stated that in many cases on-line techniques are unavailable. Then, two typical problems arise. They are (1) obtaining a representative sample from the reaction mass for analysis and (2) dealing with the "noise" in the analysis.

The nature of composition analysis requires that a side stream or discrete sample of the system be obtained prior to the analysis. Seldom, in fact, is it possible to obtain a measure of composition in the reactor proper. The patent literature abounds with devices designed to provide a discrete sample or uniform side stream from a reaction vessel. The problem in some cases is complicated by other factors, including the need to preclude exposure of the sample to air, changing composition during the sampling process due to reaction, or the physical character of the reactants which make sampling difficult. Also, the ideal sampler will ensure that each new sample is unaffected by previous samples, i.e., that the sample device is adequately purged between samples. In some cases, the measurement technique requires very tight control of or compensation for temperature or pressure within the device. All of these factors must be taken into consideration when designing or specifying an on-line sampling device for composition analysis. The result in many cases is that more time and effort are expended in development of the sampling method than is required to develop the analysis.

The second problem is the presence of a significant level of noise which is often found in composition analysis, whether the analysis is run on-line or off-line.

For continuous polymerizations, this uncertainty in the data can generally be handled by the combination of signal pretreatment or filtering and the natural attenuation of noise which occurs in a properly tuned controller-process system for component flow rate. For batch systems, however, the problem of analysis "noise" can be more severe because generally only a single sample is analyzed for each batch. In this case, a more sophisticated statistical approach, in which the on-line or lab analysis is routinely calibrated against a standard, or an on-line redundant instrumentation with data reconciliation may be necessary.

The above discussion highlights the difficulties encountered in composition measurement. Of even greater difficulty in many cases is the measurement of polymer composition or conversion. The same problems of sampling and noise exist in conversion measurement along with some basic analytical difficulties which are often not encountered in chemical composition measurements. For instance:

1. The prediction of polymer composition often involves the measurement of a bulk property of the entire reaction mass that changes as the polymerization progresses. Assumption of a boundary condition, such as initial monomer composition, is usually required to allow for the calculation of conversion. This assumption can be a source of significant measurement error.

2. Most techniques for polymer composition measurement require long sampling intervals, i.e., either the analysis requires a long dissolution, heating, or reaction time, or manpower or equipment limitations preclude more frequent analysis.

3. In copolymerization, the measurement of composition is subject to measurement error stemming from inaccurate knowledge of the instantaneous comonomer composition.

4. Many conversion measurement techniques require some pretreatment of the sample prior to measurement. For example, as a means of conversion measurement in some condensation polymerizations, the concentration of a specific end group in the bulk polymer is monitored, which requires isolation, then dissolution of the polymer before the analysis.

These considerations have dealt primarily with the difficulty of obtaining reliable measurements of composition and conversion. Additionally, from a process control standpoint, these measurements may present some problems that are dynamically significant compared to the dynamics of the process. Although the most likely dynamic component introduced by these measurements is deadtime, it is conceivable that some lag may also be present. These considerations must be taken into account when designing control strategies for conversion or composition. A final important consideration in this area is that almost all of the measurement techniques for composition are discrete in nature, as opposed to the continuous temperature, pressure, and flow measurements. Consequently, the use of continuous pneumatic or electronic controllers is essentially precluded and the control problem enters the realm of sampled-data (digital) control.

6.5 Measurement of Molecular Weight and Molecular Weight Distribution

Measurement of molecular weight (MW) and molecular weight distribution (MWD) may be considered a step closer to the desired control objective of providing regulatory control of important end-use properties of the polymer. The difficulties encountered in the provision of these measurements are also of a higher order. The typical industrial approaches for molecular weight measurement may be classified in order of frequency of application as:

1. Inferential measurement. This technique is the most common in practice. Where molecular weight or MWD is known to be a key consideration in the utility of the final product, the effects of other process or polymer variables on MW are defined. The rigor of the modeling effort varies from sophisticated on-line adaptive modeling methods to crude intuitive rules-of-thumb, but what is obtained is a "model" which relates MW or MWD to variables which can be effected in the process. A common example is the regulation of monomer conversion in continuous or batch systems to obtain the desired "molecular weight." In this mode of measurement, the actual state is not observed, but it is estimated from other measurements. Most often, the control strategy is open loop: A change in the setpoint of the measured variables is made after off-line tests have shown an error in the MW or MWD.

2. Off-line tests. Occasionally, quality control laboratories are equipped with instruments which indicate MW or MWD directly. The most common technique used for this purpose is gel permeation chromatography (GPC). Very infrequent analysis is the rule; generally, these tests are run when a problem is known to exist. Again, the more common approach is to measure a polymer property which can be related to MW or MWD on a routine basis and use these results to make setpoint corrections in the operation. These routine tests may, in fact, be the specification tests for release of the finished polymer for sale. Common among these tests is a measure of polymer viscosity, which may then be related upward to MW or MWD, and downward to process variables.

3. On-line measurement. This approach is least common in industry, but applications of on-line measurement of MWD by GPC have been reported [5]. A slight variance of this approach is to include in the process an on-line instrument to measure a property of the finished polymer, such as melt viscosity, which may be directly related to MW or MWD.

The measurement problems in the area of MW or MWD are similar to those discussed above. Deadtimes and measurement inaccuracies may be greater than for composition measurement. Adding to the difficulty is the fact that it is uncommon that a single simple relationship between a manipulated process input and MW can be identified. More commonly, a change in polymer MW is effected by the simultaneous adjustment of several process variables.

6.6 State Estimation

Often only some of the variables of interest (states) of a polymerization system are measured. As noted earlier, these are most often process variables such as temperatures, pressures, and flow rates. Other process states, such as compositions, and product states, such as MW, MWD, copolymer composition, etc., are most often unavailable. If a reasonably accurate mathematical model of the polymerization process is available, then it may be possible to estimate those states which are not measured, based on the values of those states which are. There are a variety of estimation schemes ranging from the more or less ad hoc approach mentioned above, to observers and estimators which have desirable convergence properties. Perhaps the most useful of these is the Kalman filter.

The Kalman filter is an optimal estimator for the estimation of the states of a dynamic system from a set of measurements which is a subset of the set of states (or linear combinations of states). As such, it can be used for noise filtering, estimation of unmeasured states, and prediction of future values of states. The derivation of the Kalman filtering equations can be found elsewhere [5,6]. The discrete Kalman filtering equations are given as follows:

At the measurement time (measurement update):

$$\hat{x}(n) = \bar{x}(n) + M(n)H^T(HM(n)H^T + R_v)^{-1}(\bar{y}(n) - H\bar{x}(n)) \quad (6.1)$$

$$P(n) = M(n) - M(n)H^T(HM(n)H^T + R_v)^{-1}HM(n) \quad (6.2)$$

Between measurements (time update):

$$\bar{x}(n + 1) = \Phi\hat{x}(n) + \beta u(n) \quad (6.3)$$

$$M(n + 1) = \Phi P(n)\Phi^T + R_w \quad (6.4)$$

The system is represented as:

$$x(n + 1) = \Phi x(n) + \beta u(n) \quad (6.5)$$

$$y(n) = Hx(n) \quad (6.6)$$

The vector of states is x. The vector of measurements is y. P is the covariance matrix defining the quality of the estimates at time n. R_v is the covariance matrix for the process measurements, and $\mathbf{R}_w$ is the covariance matrix for the process model. The system model above is a discretized approximation to the mathematical model of the process (in this case, the mass and energy balances describing the polymerization system). The discretization is done by taking the z-transform of the mass and energy balance equations linearized around the current operating point.

It can be seen from Eq. (6.1) that the estimate of $\mathbf{x}$ at time n $[\mathbf{x}(n)]$ is made up of a model prediction and a correction term based on the available measurements. Thus, the estimates of all of the states are corrected based on the error in estimating those states which are measured. The Kalman filter has been shown to converge to correct estimates for a wide variety of measurements. It has been used to estimate MW and MWD based on temperature and conversion measurements, with or without low-frequency GPC measurements of MWD [4,7,8]. The Kalman filter can be extended to nonlinear systems (in the form of the extended Kalman filter) by updating the approximate discrete linear model [Eq. (6.5)] against a rigorous nonlinear model at each sampling point.

6.7 Summary

What then are the key considerations from a process control point of view in regard to measurement in polymerization systems? We have seen that there are really several levels to the control problem with increasing degrees of measurement difficulty. At the highest level is the overall control objective to regulate polymer quality to a point which will provide the customer with the degree of performance and uniformity that is required in his process. Most commonly, a unique measurement which addresses this desire is not available. This desire is typically addressed by setting a release specification often related to MW or MWD, which again is not often measured on-line. Changes in the desired MW are transmitted to the plant by changing conversion, composition, temperature setpoints. And finally, the regulation of the plant requires control of temperature, pressure, and flow.

The unique measurement challenges in polymerization then may be summarized:

1. vague control and measurement objectives;
2. difficult measurement sampling;
3. discrete data;
4. large deadtimes and sampling intervals;
5. off-line analyses;
6. large measurement errors and high noise levels.

References

1. Ernest O. Doebelin, *Measurement Systems Application and Design*, 3rd ed., McGraw-Hill, New York, 1983.

2. F. J. Schork and W. H. Ray, "On-Line Monitoring of Emulsion Polymerization Reactor Dynamics," in *Emulsion Polymers and Emulsion Polymerization* (D. R. Bassett and A. E. Hamielec, eds.), American Chemical Society, Washington, D.C., 1981.
3. F. J. Schork and W. H. Ray, "On-Line Measurement of Surface Tension and Density With Applications to Emulsion Polymerization," *J. Appl. Polym. Sci.*, *28*, 407 (1983).
4. V. Gonzalez, T. W. Taylor, and K. F. Jensen, "On-Line Estimation of Molecular Weight Distributions in Methyl Methacrylate Polymerization," *Proceedings of the 1986 American Control Conference*, Seattle, June 1986.
5. Frank L. Lewis, *Optimal Estimation*, Wiley, New York, 1886.
6. W. H. Ray, *Advanced Process Control*, McGraw-Hill, New York, 1981.
7. M. F. Ellis, V. Gonzalez, T. W. Taylor, and K. F. Jensen, "Estimation of Molecular Weight Distribution in Methyl Methacrylate Solution Polymerization," *Proceeding of the 1987 American Control Conference*, Minneapolis, June 1987.
8. A. K. Adebekun and F. J. Schork, "Continuous Solution Polymerization Reactor Control. 2. Estimation and Nonlinear Reference Control During Methylmethacrylate Polymerization," *I & EC Research*, *28*, 1846–1861 (1989).

7

Process Modeling and Identification

In the context of control, a process model is a set of one or more differential equations that relate the dependent (or controlled) variables and the independent variable (time) to the forcing functions (manipulated and disturbance variables). In many examples of practical interest, these models turn out to be nonlinear. The process models can be developed by writing out the phenomenological equations, involving unsteady-state mass, energy, and momentum balances, or by experimental testing in the plant. Polymerization reactors often give rise to models that are quite complex. The complexities arise due to strong nonlinearities of the rate constants and the exothermicities of the reactions. The nonlinearities present difficulties to the linear controllers and the exothermicity, if present, can lead to open-loop unstable behavior requiring strict vigilance for reasons of safety. Some progress has been made in the area of nonlinear control system design as we will see in a later chapter, but industrial polymerization reactors for the most part are operated under the command of linear control strategies.

Linear control strategies are those that are implicitly or explicitly based on linear process models. Linear process models can be derived from nonlinear differential equations by linearizing them around a steady-state operating point. The linearized differential equations can be used in control systems design in a variety of ways. For instance, they can be Laplace transformed to derive transfer functions and used in the design of conventional PID type feedback or cascade controllers and lead–lag type feedforward controllers. They can be z-transformed to determine pulse transfer functions that are used to derive these and other custom-designed algorithms for implementation on digital control computers. They can also be used to derive state-space models that are the building blocks of optimal control theory. Finally, they can lead to step or impulse response type models that are used in model-based control.

Two developments of the last two decades are having a significant impact on industrial operations. One involves the concept of (linear) model-based (also called model-predictive) control having constraint handling capabilities leading to optimized control. MPC has been successfully applied to multivariable plants throughout the world. MPC utilizes the step or impulse response models mentioned in the previous paragraph. The other involves PRBS (pseudo-random binary sequence) testing, a procedure that leads to impulse responses (or transfer functions) and noise models which can be used to design deterministic or stochastic controllers.

We have seen in a previous chapter how phenomenological models may be derived for polymerization systems from first principles. The focus in this chapter is primarily on experimental testing and its application to the development of approximate process models. For the most part, the techniques we describe in this chapter will lead to linear process models, although one of the methods presented can be used in nonlinear identification as well. The notion of linearity imposes certain restrictions on the testing procedures as we will see.

In an experimental test on a single-loop linear system, an input variable is subjected to a change away from its normal (steady-state) operating level and the movement of the associated controlled variable is recorded. The input–output time records can be analyzed by a variety of methods to identify the process model. Figure 7.1 shows the block diagram of a single-input single-output (SISO) linear system to illustrate the concepts. As is evident, the controlled variable is affected by changes in two types of inputs: manipulated variable and process loads. For a linear SISO system, the following equation applies:

$$Y(s) = G_p(s)U(s) + G_L(s)L(s) \tag{7.1}$$

where

Y = controlled variable,
U = manipulated variable,
L = disturbance variable,
G_p = process transfer function,
G_L = load transfer function,
s = Laplace transform operator.

In principle, a single-loop system can be affected by a number of disturbances; only one has been depicted for illustrative purposes. Also, due to the assumption of linearity, the principle of superposition applies. Thus, the response of the

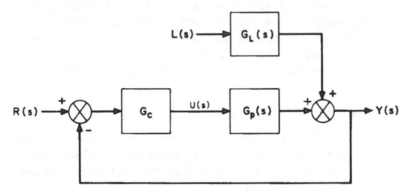

Figure 7.1 Block diagram of an SISO system.

controlled variable to the combined effect of manipulated and load variables can be evaluated by subjecting the system to a change in each variable, one at a time, and summing the effect. It should also be remembered that in some applications the process transfer function and the load transfer function are the same. In most others, they are not.

Bearing in mind that the variables Y, X, and L are in deviation form, Eq. (7.1) shows that the time records of Y and U will lead to $G_p(s)$ only if L is zero. If it is possible to vary L as well during the identification test, then the two transfer functions $G_p(s)$ and $G_L(s)$ can be evaluated according to the superposition principle. Also, if L is stochastic in nature, then the analysis of the time records will lead to the process transfer function and a noise model, the latter being a description of the effect of stochastic L upon Y, as we will see in a later section on PRBS testing.

Let us extend the foregoing concepts to multivariable process identification. Consider for illustrative purposes the following equations of a 2×2 system relating the controlled variables to the manipulated variables and a disturbance:

$$Y_1 = G_{11}U_1 + G_{12}U_2 + G_{L1}L \tag{7.2}$$

$$Y_2 = G_{21}U_1 + G_{22}U_2 + G_{L2}L \tag{7.3}$$

Again, in the absence of load disturbances, a testing procedure in which U_1 and U_2 are perturbed and Y_1 and Y_2 are recorded can be used to evaluate the four elements of the transfer function matrix **G**. The load disturbances can be accommodated according to the concepts alluded to in the last paragraph.

The organization of this chapter is as follows. We begin with a summary of state-space models, then move on to step testing, a simple method for finding approximate process models. This will be followed by pulse testing which leads to transfer functions and Bode plots. Then, we will discuss impulse response modeling; in this section we will learn how impulse and step response models of multivariable systems can be developed. Then we will take up time series analysis and PRBS testing for SISO and multivariable systems. We will end the chapter with a discussion of a random search procedure for process identification.

7.1 State-Space Models

If the mass and energy balances describing a polymerization system can be written with some degree of accuracy, the resulting set of differential equations is known as a *state-space model* of the system. This set of differential equations can be written in matrix notation as

$$\dot{x} = Ax + Bu + \Gamma L \tag{7.4}$$

Here, x is the vector of states (those variables for which the mass and energy balances have been written). $\dot{x}$ on the left-hand side of the equation represents the time derivative of the vector x. The vector u is the vector of manipulated variables or controls. The vector L is the vector of disturbances. For a linear system, the matrices A, B, and Γ are constant matrices relating x, u, and L (respectively) to the time derivative of x. For a nonlinear system, the state equation is given by

$$\dot{x} = f(x,u,L) \tag{7.5}$$

where f is a vector of nonlinear functions. The states of a polymerization system may include monomer conversion, average molecular weight, etc., as well as temperatures and concentrations. It should be obvious from Chapter 6 that not all of these states are measurable. To convert the state-space model into an input–output model necessary for control studies, a vector of measurements (outputs) y is defined as

$$y = Cx + Du \tag{7.6}$$

Here, C is a matrix describing the interrelationship between the states and the measurements, and D is a matrix relating the controls to the measurements. For a linear system, C and D are matrices of constants, and y is a subset of x, or is a vector of linear combinations of the x's, and possibly, the u's. For a nonlinear system, y is some nonlinear function of the states and controls:

$$y = h(x,u) \tag{7.7}$$

Equations (7.4), (7.6) and (7.5), (7.7) constitute a state-space model for a linear and nonlinear system, respectively. A nonlinear state-space model can be approximated by a linear model by substituting the Jacobians of f with respect to x, u, and L for **A**, **B**, and **C** respectively, in Eq. (7.4), and the Jacobians of **h** with respect to **x** and **u** for **C** and **D**, respectively, in Eq. (7.6) [1].

The linear state-space model can be converted to the more familiar Laplace domain description

$$\mathbf{y}(s) = \mathbf{G}_p(s)\mathbf{u}(s) + \mathbf{G}_d(s)\mathbf{L}(s) \tag{7.8}$$

by the relationships

$$\mathbf{G}_p(s) = \mathbf{C}(s\mathbf{I} - \mathbf{A})^{-1}\mathbf{B} \tag{7.9}$$

$$\mathbf{G}_d(s) = \mathbf{C}(s\mathbf{I} - \mathbf{A})^{-1}\mathbf{\Gamma} \tag{7.10}$$

where $\mathbf{G}_p(s)$ is the matrix of process transfer functions, $\mathbf{G}_d(s)$ is the matrix of disturbance transfer functions, and **I** is the identity matrix.

7.2   Step Testing

Step testing is one of the simplest identification procedures available. The method consists of step-forcing an input and recording the response of the associated output. The resulting data may be analyzed to determine the process model, usually in the form of a transfer function or an impulse response model. The test is usually conducted with the controllers in manual and, therefore, the resulting data yield an open-loop model. But, in principle, the test can be conducted with the controller in automatic, in which case the transient data will lead to a closed-loop model from which an approximate open-loop model can be developed if desired. Closed-loop testing is advantageous for systems that are open-loop unstable because for such systems, the feedback controller cannot be put in manual, even during the test. In the following paragraphs, we describe simple graphical procedures to fit first-order and overdamped second-order transfer function models to the step response data. This will be followed by an explanation of how an impulse response model can be developed from step response data.

To begin, consider the three typical plots resulting from an open-loop step test shown in Fig. 7.2. Note that all three responses shown indicate that the process is open-loop stable. Each has a nonminimum-phase element arising out of the presence of deadtime or inverse response. In this section, we will primarily be concerned with responses of the type shown in Figs. 7.2a and 7.2b. An optimization method of the type discussed in an ensuing section can be used to handle any of the three responses shown in Fig. 7.2, as we will see.

7.2.1   First-Order Models

The response curves of the type shown in Figs. 7.2a and 7.2b can be adequately represented by a first- or second-order model with deadtime. To fit the response

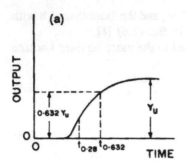

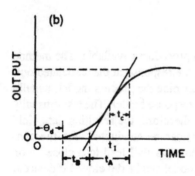

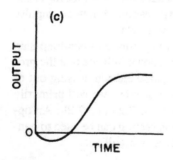

Figure 7.2 Typical plots from open-loop tests.

curve to a first-order model with deadtime, we make use of the mathematical model for this type of system given by the differential equation

$$\tau_p \frac{dY}{dt} + Y(t) = K_p X(t - \theta_d) \tag{7.11}$$

where

t = time,
K_p = process steady-state gain,
θ_d = deadtime,
τ_p = process time constant.

The analytical solution of this differential equation for the case where $X(t)$ is a step function of magnitude M is

$$Y(t) = MK_p\left(1 - \exp\left[\frac{-(t - \theta_d)}{\tau_p}\right]\right) \tag{7.12}$$

Note that $Y(t)$ and M are in deviation form. The process steady-state gain is calculated by the formula

$$K_p = \frac{Y_u}{M} \tag{7.13}$$

where Y_u is the ultimate (steady-state) value of $Y(t)$.

If the process were truly first order with deadtime, then only two pairs of values of Y and t would be needed to obtain θ_d and τ_p. Thus, if the 28% and 63.2% response values shown in Fig. 7.2A are utilized, the formulas are

$$\tau_p = 1.5(t_{0.632} - t_{0.28}) \tag{7.14}$$

and

$$\theta_d = 1.5\left(t_{0.28} - \frac{1}{3}t_{0.632}\right) \tag{7.15}$$

Another procedure is to take logarithm of Eq. (7.12), giving

$$\ln\left[1 - \frac{Y(t)}{MK_p}\right] = -\frac{t}{\tau} + \frac{\theta_d}{\tau} \tag{7.16}$$

Equation (7.16) would give a straight line on a semilog graph paper having a y intercept of θ_d/τ and a slope of $-1/\tau$ and, thus, θ_d and τ can be determined.

7.2.2 Second-Order Models

An open-loop step response curve of the type shown in Fig. 7.2b is adequately described by an overdamped second-order process model containing deadtime. High-order complex industrial processes often yield this type of responses. To fit this curve to an overdamped second-order model, we begin with a mathematical model given by the differential equation

$$\tau_1 \tau_2 \frac{d^2Y}{dt^2} + (\tau_1 + \tau_2) \frac{dY}{dt} + Y(t) = K_p X(t - \theta_d) \tag{7.17}$$

For a step change in the input of magnitude M, the solution of Eq. (7.17) is

$$Y(t) = MK_p \left[1 - \frac{\tau_1 \exp[-(t - \theta_d)/\tau_1] - \tau_2 \exp[-(t - \theta_d)/\tau_2]}{\tau_1 - \tau_2} \right]$$

$$\tag{7.18}$$

The parameters to be determined are K_p, τ_1, τ_2, and θ_d. To determine these parameters graphically [2], a tangent is drawn at the inflection point as shown in Fig. 7.2b. Its intersection at $Y = 0$ and $Y = MK_p$ leads to two equations,

$$T_A = \tau_1 \left(\frac{\tau_2}{\tau_1} \right)^{\tau_2/(\tau_2 - \tau_1)} \tag{7.19}$$

and

$$T_C = \tau_1 + \tau_2 \tag{7.20}$$

from which the two unknowns τ_1 and τ_2 can be calculated by trial and error. The deadtime can be calculated by the equation

$$\theta_d = t_I + T_C - T_A - T_B \tag{7.21}$$

and the process steady-state gain according to

$$K_p = \frac{Y_u}{M} \tag{7.22}$$

A drawback of this method is that the correct location of the inflection point is often not an easy task. In any event, the model can be inverted into the time domain and the model response plotted alongside the experimental response to check the goodness-of-fit.

7.3 Pulse Testing

In pulse testing [3, 4], a pulse of arbitrary shape is applied to the input of the process while the process is operating at steady state with the feedback controller in manual and the transient response of the output is recorded. The input–output records can be used to develop a frequency-response diagram as described is the following paragraphs.

The open-loop process transfer function is given by

$$G(s) = \frac{Y(s)}{X(s)} \tag{7.23}$$

where Y is the process output and X is the input to which the pulse is applied. The input may be a manipulated variable or a process load depending on whether process dynamics or load dynamics are to be determined. To begin, the definition of Laplace transforms is substituted into Eq. (7.23) giving

$$G(s) = \frac{\int_0^\infty Y(t)e^{-st}\,dt}{\int_0^\infty X(t)e^{-st}\,dt} \tag{7.24}$$

Substituting $j\omega$ for s in Eq. (7.24) gives

$$G(j\omega) = \frac{\int_0^\infty Y(t)e^{-j\omega t}\,dt}{\int_0^\infty X(t)e^{-j\omega t}\,dt} \tag{7.25}$$

The upper limits on the integrals may be changed to t_y and t_x, the time periods during which changes in Y and X occur. Furthermore, if $\cos(\omega t) - j\sin(\omega t)$ is substituted for $e^{-j\omega t}$ into Equation (7.25), we obtain

$$G(j\omega) = \frac{\int_0^{t_y} Y(t)\cos(\omega t)\,dt - j\int_0^{t_y} Y(t)\sin(\omega t)\,dt}{\int_0^{t_x} X(t)\cos(\omega t)\,dt - j\int_0^{t_x} X(t)\sin(\omega t)\,dt} \tag{7.26}$$

$$G(j\omega) = \frac{A - jB}{C - jD} \tag{7.27}$$

where A, B, C, and D are the four integrals in Eq. (7.26). The amplitude ratio and phase angle can now be calculated as

$$AR = |G(j\omega)| = \sqrt{\text{Re } G(j\omega) + \text{Im } G(j\omega)}$$

$$= \sqrt{\left(\frac{AC + BD}{C^2 + D^2}\right)^2 + \left(\frac{AD - BC}{C^2 + D^2}\right)^2} \tag{7.28}$$

$$\phi = \angle\, G(j\omega) = \arctan\left(\frac{\text{Im } G(j\omega)}{\text{Re } G(j\omega)}\right)$$

$$= \arctan\left(\frac{AD - BC}{AC + BD}\right) \tag{7.29}$$

The computational procedure is to choose a specific value of frequency, ω, and perform the integrations in Eq. (7.26) by a suitable numerical integration subroutine giving a single value of A, B, C, and D which, on substitution in Eqs. (7.28) and (7.29), give a point on the frequency–response diagram. The procedure is repeated for numerous values of ω and, thus, the complete frequency–response diagram can be prepared.

In conducting the pulse test, the control engineer must ascertain that the process is free of upsets in variables other than the one being pulsed throughout the test. The pulse shape, height, and duration must be selected carefully if meaningful frequency–response information is to be expected. The height of the input pulse must be such that the range of linearity is not exceeded. This necessitates several tests with pulses of various heights. Sound dynamic testing procedures suggest that tests with positive and negative pulses should be conducted. The computed Bode plots then give an indication of the nonlinearities present.

The selection of the proper duration of the input pulse is equally important. A good rule-of-thumb is to select a pulse width that is smaller than half the smallest time constant of interest [5]. If the process dynamics are completely unknown, it may take a few trials to establish a suitable pulse width. If the pulse width is too small for a pulse of given height, the system is disturbed very little, and, therefore, it becomes difficult to distinguish between real process output and process noise.

7.4 Impulse Response Modeling

In model-predictive methods such as dynamic matrix control, model algorithmic control (commercially known as Idcom), and internal model control, an impulse or step response model is used as a basis for controller design. The descripter "predictive" in the name is meant to indicate that predicted future values of controlled and manipulated variables are used in the design procedure. Owing to constraint handling, deadtime compensation, and interaction compensation capabilities of these techniques, they have been successfully applied to a number of complex multivariable plants throughout the world. Some examples include crude towers, fluid catalytic crackers, petrochemical distillation towers, blending systems, and chemical reactors.

In this section, we show how an impulse or step response model may be derived from experimental testing in a plant. The derivation assumes that the reader is familiar with z-transforms and sampled data control concepts. We also assume that the plant is open-loop stable. For open-loop unstable systems, stability can be achieved by applying conventional feedback and then identification may be applied to the resulting system. For an integrating process, the output(s) may be differenced to yield open-loop stability [6]. The identification procedure described in the following subsections also requires an understanding of an impulse or step response model. Such a model may be derived as follows.

7.4.1 Pulse Transfer Functions in Terms of Impulse or Step Responses

The open-loop transfer function of a process can be represented by the expression

$$G(z) = \frac{Y(z)}{U(z)} \tag{7.30}$$

For a single-unit step function,

$$U(z) = \frac{1}{1 - z^{-1}} \tag{7.31}$$

Substitution of $U(z)$ in Eq. (7.30) gives

$$G(z) = (1 - z^{-1})Y(z)|_{step} \tag{7.32}$$

If, on the other hand, the input were a unit impulse function, $U(z) = 1$ then Eq. (7.30) would give

$$G(z) = Y(z)|_{impulse} \tag{7.33}$$

Equating Eqs. (7.32) and (7.33), we find that

$$Y(z)|_{impulse} = (1 - z^{-1})Y(z)|_{step} \tag{7.34}$$

Now, the response of a process to a single-unit step change in input shown in Fig. 7.3 may be described according to

$$Y(z)|_{step} = a_1 z^{-1} + a_2 z^{-2} + \cdots \tag{7.35}$$

From Eqs. (7.33), (7.34), and (7.35), we obtain

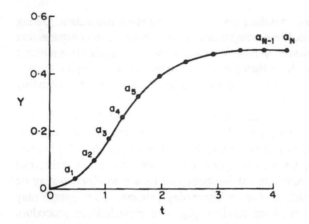

Figure 7.3 Process response to a unit step change in input.

$$G(z) = Y(z)|_{\text{impulse}} = (1 - z^{-1})(a_1 z^{-1} + a_2 z^{-2} + \cdots)$$

$$= a_1 z^{-1} + (a_2 - a_1)z^{-2} + \cdots$$

$$= h_1 z^{-1} + h_2 z^{-2} + \cdots \tag{7.36}$$

The numbers in the sequence (h_i, $i = 1,2,...$) are called *impulse response coefficients* and a system represented in the form as Eq. (7.36) is said to have a model of the impulse response (IR) type. Similarly, the numbers in the sequence (a_i, $i = 1,2,...$) are called *step response coefficients* and a system represented in terms of this sequence is said to have a model of the step response type.

The open-loop response given in Eqs. (7.35) or (7.36) is based on a single step input. In impulse response modeling, a train of randomly generated up and down step changes are introduced at the input, and, therefore, a representation is needed to relate this type of input to the process output. The response of a process to a train of step changes may simply be obtained by the superposition principle, which states that the response to a sum of several functions is equal to the sum of the responses to the individual functions acting separately. A train of step changes can be represented as

$$U(z) = U_0 z^0 + U_1 z^{-1} + U_2 z^{-2} + \cdots \tag{7.37}$$

where U_0 is the magnitude of the step input occurring at $t = 0$, U_1 is the magnitude of the step input occurring at $t = T$, etc., where T is the sampling period. Now let us rearrange Eq. (7.30) and write it as

$$Y(z) = G(z)U(z) \tag{7.38}$$

Substituting for $G(z)$ and $U(z)$ from Eqs. (7.36) and (7.37), respectively, in Eq. (7.38), we get

$$Y(z) = (h_1 z^{-1} + h_2 z^{-2} + \cdots)(U_0 z^0 + U_1 z^{-1} + U_2 z^{-2} + \cdots)$$

$$= (h_1 U_0) z^{-1} + (h_2 U_0 + h_1 U_1) z^{-2} + \cdots + (h_n U_0 + h_{n-1} U_1$$

$$+ \cdots h_1 U_{n-1}) z^{-n} + \cdots \tag{7.39}$$

$Y(z)$ can equivalently be written as

$$\hat{Y}(z) = \hat{Y}_1 z^{-1} + \hat{Y}_2 z^{-2} + \hat{Y}_3 z^{-3} + \cdots \tag{7.40}$$

where the symbol $\hat{\ }$ indicates a predicted value. Comparing Eqs. (7.39) and (7.40) reveals that

$$\hat{Y}_{k+1} = h_1 U_k + h_2 U_{k-1} + \cdots + h_k U_1 + h_{k+1} U_0 \tag{7.41}$$

or, in alternate form,

$$\hat{Y}_{k+1} = \sum_{i=1}^{N} h_i U_{k+1-i} \tag{7.42}$$

Note that $U_i = 0$ for $i < 0$. Equation (7.42) enables us to predict the response of Y to a train of past step inputs.

7.4.2 Impulse Response Modeling of SISO Systems

Suppose it is desired to model an SISO plant that is disturbed by a single measurable disturbance. The goal of the modeling exercise is to develop an impulse response model relating the output, Y, to the manipulated variable M and the disturbance variable L. Toward this goal, a series of step changes are introduced in M and L and the response of Y and the inputs are recorded. These inputs must be uncorrelated with respect to each other, a requirement which may be met by employing some random procedure for generating them. The direction of the step change should not be switched any more often than about half the estimated time constant of the open-loop system, in accordance with the guidelines on duration of a pulse in pulse testing. Now, the output Y is related to the two inputs according to

$$Y_{k+1}^M = h_1^M M_k + h_2^M M_{k-1} + \cdots + h_k^M M_1 + h_{k+1}^M M_0 \tag{7.43}$$

and

$$Y_{k+1}^L = h_1^L L_k + h_2^L L_{k-1} + \cdots + h_k^L L_1 + h_{k+1}^L M_0 \tag{7.44}$$

with

$$Y_{k+1} = Y_{k+1}^M + Y_{k+1}^L \tag{7.45}$$

The output Y at the various sampling instants can be evaluated according to

$$Y_1 = h_1^M M_0 + h_1^L L_0$$

$$Y_2 = h_1^M M_1 + h_2^M M_0 + h_1^L L_1 + h_2^L L_0$$

$$Y_3 = h_1^M M_2 + h_2^M M_1 + h_3^M M_0 + h_1^L L_2 + h_2^L L_1 + h_3^L L_0 \qquad (7.46)$$

and so on.

For an illustrative example where a total of six impulse response coefficients are to be identified, these equations may be organized in the matrix form as

$$
\begin{bmatrix} Y_1 \\ Y_2 \\ Y_3 \\ Y_4 \\ Y_5 \\ Y_6 \end{bmatrix} =
\begin{bmatrix}
M_0 & 0 & 0 & 0 & 0 & 0 & L_0 & 0 & 0 & 0 & 0 & 0 \\
M_1 & M_0 & 0 & 0 & 0 & 0 & L_1 & L_0 & 0 & 0 & 0 & 0 \\
M_2 & M_1 & M_0 & 0 & 0 & 0 & L_2 & L_1 & L_0 & 0 & 0 & 0 \\
M_3 & M_2 & M_1 & M_0 & 0 & 0 & L_3 & L_2 & L_1 & L_0 & 0 & 0 \\
M_4 & M_3 & M_2 & M_1 & M_0 & 0 & L_4 & L_3 & L_2 & L_1 & L_0 & 0 \\
M_5 & M_4 & M_3 & M_2 & M_1 & M_0 & L_5 & L_4 & L_3 & L_2 & L_1 & L_0
\end{bmatrix} =
\begin{bmatrix} h_1^M \\ h_2^M \\ \vdots \\ h_6^M \\ h_1^L \\ h_2^L \\ \vdots \\ h_6^L \end{bmatrix}
$$

$$(7.47)$$

or

$$\mathbf{Y} = \mathbf{X}\mathbf{H} \qquad (7.48)$$

Therefore

$$\mathbf{H} = \mathbf{X}^t \mathbf{Y} \qquad (7.49)$$

where $\mathbf{X}^t$ is the pseudoinverse, $(\mathbf{X}^T\mathbf{X})^{-1}\mathbf{X}^T$. Note that Eq. (7.49) is also the least squares solution. The impulse response coefficients, h_i^M are used to design feedback controllers, whereas h_i^L are used to design feedforward controllers.

7.4.3 Impulse Response Modeling of MIMO Systems

The extension of the foregoing procedure to multivariable systems is straightforward. Suppose we wish to identify the dynamics of a 2×2 process with one disturbance variable which must be considered. In this instance, we would vary the two manipulated variables and the disturbance variable in a stepwise fashion and record the responses of the two outputs along with the inputs. Suppose each transfer function may be adequately described in terms of three impulse response coefficients. Then, the system equations for the first output may be written down as

$$Y_1^1 = h_{11}^1 M_1^0 + h_{21}^1 M_2^0 + h_L^1 L^0$$

$$Y_1^2 = h_{11}^1 M_1^1 + h_{11}^2 M_1^0 + h_{21}^1 M_2^1 + h_{21}^2 M_2^0 + h_L^1 L^1 + h_L^2 L^0 \qquad (7.50)$$

$$\vdots$$

$$Y_1^9 = h_{11}^1 M_1^8 + \cdots + h_{11}^9 M_1^0 + h_{21}^1 M_2^8 + \cdots + h_{21}^9 M_2^0 + h_L^1 L^8$$
$$\quad + \cdots + h_L^9 L^0$$

Similar equations can be written down for Y_2. These equations can be compactly written as

$$Y_1 = X_1 H_1 \qquad (7.51)$$

where

Y_1 is of dimension 9×1,
H is of dimension 27×1,
X_1 is of dimension 9×27.

The set in Eq. (7.51) has 9 unknowns and they can be computed according to

$$H_1 = X_1^1 Y_1 \qquad (7.52)$$

A similar procedure may be employed to calculate h_{12}^i and h_{22}^i according to

$$H_2 = X_2^1 Y_2 \qquad (7.53)$$

It should be clear that as the dimension of the system and the number of impulse response coefficients increase, the dimension of these matrices increases dramatically. Commercial packages such as the one marketed by DMC Corporation efficiently handle these computations.

7.5 Time Series Analysis

A time series is a set of observations of a variable at discrete equispaced intervals of time, denoted by the sampling period [6]. In the context of an SISO (single-input single-output) system, the process is deemed to be affected by a manipulated variable and a set of one or more disturbances. The combined effect of the disturbances on the output is described in terms of a stochastic noise model. The object of time series analysis is to identify the combined transfer function–noise model of a linear system given a set of observations of input and output by statistical methodology. The method requires that the input and noise must be statistically independent. This requirement often precludes the use of normal operating data to identify models. Rather, the process is deliberately perturbed by applying an input that is generated by some random procedure, and the resulting output data along with the input data are analyzed to develop the combined models as we will see. Computer programs are available in the market for carrying out this analysis (e.g., IDSA and MATLAB), so the burden on the reader is only to understand the underlying concepts. In the following subsections we take up SISO systems first and then show how the technique may be extended to multivariable systems. Due to the scope of this work, the treatment is necessarily brief, but the interested reader may refer to specialized books on identification and forecasting such as the one by Box and Jenkins [6] for details.

7.5.1 Some Definitions

The following definitions will be found useful in time series analysis.

z-Transform of Delayed Functions

Although we will basically be working in the time domain, the examination of functions of time in the z-domain would none the less be useful to the reader already familiar with z-transforms. Recall that the z-transform of a function of time is defined as

$$z\{f(t)\} \equiv F(z) = \sum_{n=0}^{\infty} f(nT)z^{-n} \tag{7.54}$$

where T is the sampling period. Also, the z-transform of a function of time that is delayed by k sampling instants is given by

$$z\{f(t - kT)\} = z^{-k}F(z) \tag{7.55}$$

Consequently,

$$z^{-1}[z^{-1}z\{f(t)\}] = f(t - T) \tag{7.56}$$

and

$$z^{-1}[(1 - z^{-1})z\{f(t)\}] = f(t) - f(t - T) \tag{7.57}$$

Thus, the term z^{-1} can be used to relate functions that are delayed.

Backward Shift Operator

Let m_t be the value of the variable in a time series at the current time instant t and its values at the previous sampling instants be denoted as $m_{t-1}, m_{t-2}, \ldots$. Then, the *backward shift operator* B is defined by the equation

$$Bm_t = m_{t-1} \tag{7.58}$$

and, therefore,

$$B^k m_t = m_{t-k} \tag{7.59}$$

The *forward shift operator* $P = B^{-1}$ follows the relationship

$$PM_t = M_{t+1} \tag{7.60}$$

and

$$P^k M_t = M_{t+k} \tag{7.61}$$

Using these definitions, the backward difference operator ∇ may be written as

$$\nabla M_t = M_t - M_{t-1} = (1 - B)M_t \qquad . \tag{7.62}$$

Autocovariance and Autocorrelation Coefficients

The covariance between two observations Z_t and Z_{t+k} separated by k sampling instants is called the autocovariance at lag k. For a time series $Z_1, Z_2, ..., Z_N$ of finite N observations, the autocovariance can only be estimated. The estimate, called the autocorrelation function, may be computed according to

$$r_{zz}(k) = \frac{C_{zz}(k)}{C_{zz}(0)} \tag{7.63}$$

where

$$C_{zz}(k) = \frac{1}{N} \sum_{t=1}^{N-k} (Z_t - \bar{Z})(Z_{t+K} - \bar{Z}), \quad K = 0, 1, ... \tag{7.64}$$

and $\bar{Z}$ is the mean of the time series

These equations apply to a time series having a stationary mean. A time series Z_t not having a stationary mean may be differenced d times to yield a new time series Z_t where $Z_t = \nabla^d Z_t$, which does have a stationary mean.

Partial Autocorrelation Function

A partial autocorrelation function of a stochastic time series $\hat{\phi}_{kj}$ is related to the autocorrelation function r_j according to

$$r_j = \hat{\phi}_{k1} r_{j-1} + \hat{\phi}_{k2} r_{j-2} + \cdots + \hat{\phi}_{k(k-1)} r_{J-k+1} + \hat{\phi}_{kk} r_{j-k} \tag{7.65}$$

with $J = 1, 2, ..., k$. These equations give

$$\hat{\phi}_{ll} = \frac{r_l - \sum_{j=1}^{l-1} \hat{\phi}_{l-1,j} r_{l-j}}{1 - \sum_{j=1}^{l-1} \hat{\phi}_{l-1,j} r_j}, \quad l = 2, 3, ..., L \tag{7.66}$$

$$= r_1 \quad l = 1$$

where

$$\hat{\phi}_{lj} = \hat{\phi}_{l-1,j} - \hat{\phi}_{ll}\,\hat{\phi}_{l-1,l-j}, \; j = 1, 2, ..., l-1 \tag{7.67}$$

The autocorrelation function together with the partial autocorrelation function help identify the structure of the noise model.

Cross-Covariance and Cross-Correlation Function

An estimate $C_{xy}(k)$ of the cross-covariance coefficient at lag k may be calculated according to

$$C_{xy}(k) = \begin{cases} \dfrac{1}{N} \displaystyle\sum_{t=1}^{N-k} (x_t - \bar{x})(y_{t+k} - \bar{y}), & k = 0, 1, 2, ... \\[4mm] \dfrac{1}{N} \displaystyle\sum_{t=1}^{N+k} (y_t - \bar{y})(x_{t-k} - \bar{x}), & k = 0, -1, -2, ... \end{cases} \tag{7.68}$$

Similarly, the estimate of the cross-correlation function at lag k, $r_{xy}(k)$, may be determined according to

$$r_{xy}(k) = \frac{C_{xy}(k)}{S_x S_y} \tag{7.69}$$

where

$$S_x = \sqrt{C_{xx}(0)} \tag{7.70}$$

and

$$S_y = \sqrt{C_{yy}(0)} \tag{7.71}$$

Noise Models

A series of random variables a_t, a_{t-1}, a_{t-2}, ... having zero mean and variance σ_a^2 is called white noise. A variety of noise models to represent the load variables may be generated by passing white noise through a series of one or more suitably selected filters. In the context of process control, two types of noise models representing disturbance are needed. In one, the series would have a fixed mean, whereas in the other, it would not vary about a fixed mean. Such a time series can be generated by passing white noise thorough a series of three filters as shown in Fig. 7.4, giving

$$\phi(B)W_t = \theta(B)a_t \tag{7.72}$$

with

$$W_t = \nabla^d Z_t \tag{7.73}$$

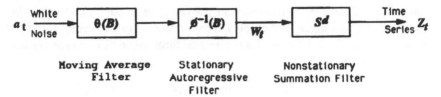

Figure 7.4 Block diagram of ARIMA model.

where

$$\phi(B) = 1 - \phi_1 B - \phi_2 B^2 - \cdots - \phi_p B^p \qquad (7.74)$$

$$\theta(B) = 1 - \theta_1 B - \theta_2 B^2 - \cdots - \theta_q B^q \qquad (7.75)$$

and $d = 0$, 1, or at most 2. Equation (7.72) gives the description of the so-called ARIMA (autoregressive integrated moving average) model of order (P, d, q), where P and Q are usually less than or equal to 2 or so.

7.5.2 Combined Transfer Function–Noise Model

In the context of time series analysis, a pulse transfer function description is appropriate for relating input–output data. Recall from the study of sampled data control concepts that the pulse transfer function of an SISO system relating the output and input is given by

$$\frac{Y(z)}{X(z)} = z\{G_{h0}(s)G_p(s)\} \qquad (7.76)$$

where

$G_p(s)$ = Laplace domain transfer function,
$G_{h0}(s)$ = transfer function of zero-order hold.

For the particular choice of an overdamped second-order with deadtime model for the plant, the application of Eq. (7.76) gives

$$\frac{Y(z)}{X(z)} = z\left\{ \frac{1 - e^{-sT}}{s} \cdot \frac{K_p e^{-\theta_d s}}{(\tau_1 s + 1)(\tau_2 s + 1)} \right\} \qquad (7.77)$$

Cross-multiplication and inversion gives

$$Y_t = a_1 Y_{t-1} - a_2 Y_{t-2} + b_1 X_{t-N-1} + b_2 X_{t-N-2} + b_3 X_{t-N-3} \qquad (7.78)$$

where

$$a_1, a_2, b_1, b_2, b_3 = f(K_p, \theta_d, \tau_1, \tau_2, T)$$

Equation (7.78) may be used to calculate the values of an output at the sampling instant t from past inputs and outputs. This equation can be readily employed in time series analysis, although a more general form would be more appropriate. Such an equation is

$$(1 - \delta_1 B - \cdots - \delta_r B^r)Y_t = (\omega_0 - \omega_1 B - \cdots \omega_s B^s)X_{t-b}$$

$$= (\omega_0 B^b - \omega_1 B^{B+1} - \cdots \omega_s B^{S+b})X_t$$

(7.79)

Equation (7.79) can be more compactly written as

$$\delta(B)Y_t = \omega(B)B^b X_t \tag{7.80}$$

or

$$Y_t = \delta^{-1}(B) \ \omega(B) \ B^b X_t \tag{7.81}$$

Recall from our previous discussion that the relationship between an input and an output may be described in terms of an impulse response model, i.e.,

$$Y_t = h(B)X_t \tag{7.82}$$

$$= h_0 X_t + h_1 X_{t-1} + h_2 X_{t-2} + \cdots$$

Therefore, a comparison of Eqs. (7.81) and (7.82) reveals that

$$h(B) = \delta^{-1}(B)\omega(B)B^b \tag{7.83}$$

In the light of Eqs. (7.72) and (7.81), the combined transfer function noise model can be written as

$$Y_t = \delta^{-1}(B)\omega(B)X_{t-b} + \phi^{-1}(B)\theta(B)a_t \tag{7.84}$$

The process is perturbed by an input time series X_t that is statistically uncorrelated with the noise N_t and the observations of the output time series Y_t are recorded. To ensure statistical independence, a random procedure may be employed to generate X_t. The use of white noise or a PRBS (pseudo-random binary sequence) signal for this purpose has been suggested. A PRBS signal consists of a train of pulses of varying width but constant height. A toss of a coin or tables of random numbers can be used to decide whether the direction of the pulse should be switched or not. The switching period is usually much larger than the sampling period. Based on the results from pulse testing, a switching period of about half the open-loop system time constant may be appropriate. The magnitude of the pulse should be sufficiently large to distinguish the output signal from noise but not so large as to drive the system out of the linear range of operation. An illustration PRBS signal is shown in Fig. 7.5.

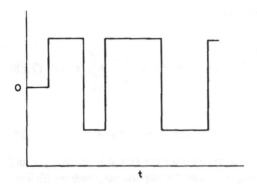

Figure 7.5 PRBS signal.

In general, the time series X_t and Y_t may not be stationary. For example, liquid level systems connected to cascaded flow controllers exhibit integrating behavior. The statistical method of analysis requires time series having constant means. Stationarity can be achieved by differencing the time series d times. Thus,

$$x_t = \nabla^d X_t \tag{7.85}$$

$$y_t = \nabla^d Y_t \tag{7.86}$$

where $d = 0$, 1, or 2, usually suffice. Stationarity may be assumed to have been achieved if for a particular choice of d, the estimated autocorrelations and cross-correlations $\gamma_{xx}(k)$, $\gamma_{yy}(k)$, and $\gamma_{xy}(k)$ of x_t and y_t damp out quickly.

The combined transfer function noise model may then be written as

$$
\begin{aligned}
y_t &= \delta^{-1}(B)\omega(B)x_{t-b} + n_t \\
&= h(B)x_t + n_t \\
&= h_0 x_t + h_1 x_{t-1} + h_2 x_{t-2} + \cdots + n_t
\end{aligned}
\tag{7.87}
$$

where $y_t = \nabla^d Y_t$, $x_t = \nabla^d X_t$, and $n_t = \nabla^d N_t$ are stationary time series. The objective of time series analysis is to analyze the input output data (x_t, y_t) to determine the parameters of the transfer function and the noise model. The step-by-step procedure is as follows.

Step 1. Determine preliminary estimates of the impulse response coefficients, h_j. There are two methods for finding, h_j.

Method I. Multiply both sides of Eq. (7.87) by x_{t-k} $(k \geq 0)$ and take expectations giving

$$\frac{1}{N} \sum_{t=1}^{N-k} x_{t-k}y_t = \frac{1}{N} \sum_{t=1}^{N-k} h_0 x_{t-k} x_t$$

$$+ \frac{1}{N} \sum_{t=1}^{N-k} h_1 x_{t-k} x_{t-1} \sum_{t=1}^{N-k} h_2 x_{t-k} x_{t-2} + \cdots + \frac{1}{N} \sum_{t=1}^{N-k} x_{t-k} n_t \quad (7.88)$$

or

$$C_{xy}(k) = h_0 C_{xx}(k) + h_1 C_{xx}(k-1) + h_2 C_{xx}(k-2) + \cdots \quad (7.89)$$

for $k = 0, 1, 2, \ldots$ where $C_{xn}(k)$ is zero under the assumption that x_{t-k} and n_t are uncorrelated for all k. Noting also that the impulse response coefficients vanish beyond $k = K$, the first $K + 1$ equations of the set given in Eq. (7.89) may be written as

$$\begin{bmatrix} C_{xy}(0) \\ C_{xy}(1) \\ \vdots \\ C_{xy}(K) \end{bmatrix} = \begin{bmatrix} C_{xx}(0) & C_{xx}(1) & \cdots & C_{xx}(K) \\ C_{xx}(1) & C_{xx}(0) & \cdots & C_{xx}(K-1) \\ & & \cdots & \vdots \\ C_{xx}(K) & C_{xx}(K-1) & \cdots & C_{xx}(0) \end{bmatrix} \begin{bmatrix} h_0 \\ h_1 \\ \vdots \\ h_K \end{bmatrix} \quad (7.90)$$

These equations can be solved for h (provided an estimate of the open-loop setting time is available from which K may be computed). Note that in writing down the equations in the foregoing set, we have made use of the fact that $C_{xx}(k) = C_{xx}(-k)$. In general, $C_{xy}(k) \neq C_{xy}(-k)$.

Method II. If the input x_t is other than white noise, a transformation may be performed so that the transformed series can be closely approximated as white noise. This procedure is called prewhitening. To apply this procedure, multiply both sides of Eq. (7.87) by $\phi_x(B)\theta_x^{-1}(B)$, giving

$$\phi_x(B)\theta_x^{-1}(B)y_t = h(B)\phi_x(B)\theta_x^{-1}(B)x_t + \phi_x(B)\theta_x^{-1}(B)n_t \quad (7.91)$$

or

$$\beta_t = h(B)\alpha_t + \epsilon_t \quad (7.92)$$

This procedure converts the correlated input series x_t into the uncorrelated white noise series α_t. These two time series are related according to

$$\phi_x(B)\theta_x^{-1}(B)x_t = \alpha_t \quad (7.93)$$

or

$$x_t = \phi_{1x}x_{t-1} + \cdots + \phi_{px}x_{t-p} + \alpha_t - \theta_{1x}\alpha_{t-1} - \cdots - \theta_{qx}\alpha_{t-q} \quad (7.94)$$

The next step is to determine estimates of the parameters ϕ_x and θ_x. To do this, we determine the autocorrelation functions $r_{xx}(k)$ and refer to appropriate figures and tables which give estimates of the parameters ϕ_x and θ_x as a function of $r_{xx}(1)$, $r_{xx}(2)$, etc., for various types of stochastic models.

Finally, both sides of Eq. (7.92) are multiplied by α_{t-k} and expectations are taken giving

$$\frac{1}{N} \sum_{t=1}^{N-k} \alpha_{t-k}\beta_t = h(B) \frac{1}{N} \sum_{t=1}^{N-k} \alpha_{t-k}\alpha_t + \frac{1}{N} \sum_{t=1}^{N-k} \alpha_{t-k}\epsilon_t \qquad (7.95)$$

or

$$C_{\alpha\beta}(k) = h_k \sigma_\alpha^2 \qquad (7.96)$$

Estimates of h_k may now be obtained according to

$$h_k = \frac{C_{\alpha\beta}(k)}{\sigma^2} \qquad (7.97)$$

where σ_α^2 is the variance of the white noise series α_t. It can be estimated from the sum of squares of the α's.

Step 2. With the impulse response coefficients determined in Step 1, estimate the orders of the polynomials r and s and the time delay b of the process transfer function. These guesses may be made in the light of the following observations:

a.   There will be b zero values of h_1, h_2, ..., h_{b-1}.
b.   A further $s - r + 1$ values, h_b, h_{b+1}, ..., h_{b+s-r}, will exhibit no fixed pattern (no such values will occur if $s < r$).
c.   The values of $h_{b+s-r+1}$, $h_{b+s-r+2}$, ... will follow a pattern of an rth-order difference equation having r starting values h_{b+s}, ..., $h_{b+s-r+1}$.

Step 3. Knowing h_i along with b, r, and s, the parameters δ and ω of the transfer function model may be found from the equations in Table 7.1.

Step 4. Estimate the noise series. For this purpose, we begin by solving Eq. (7.87) for n_t, giving

$$n_t = y_t - h(B)x_t \qquad (7.98)$$

Recall that $n_t = \nabla^d N_t$ is a stationary time series. Assuming an ARIMA model, the noise series may be written as

$$(1 - \phi_1 B - \cdots - \phi_p B^p)n_t = (1 - \theta_1 B - \cdots - \theta_q B^q)a_t \qquad (7.99)$$

or

$$n_t = \phi_1 n_{t-1} + \cdots + \phi_p n_{t-p} + a_t - \theta_1 a_{t-1} - \cdots - \theta_q a_{t-q} \qquad (7.100)$$

Table 7.1 Impulse response functions for transfer function models of the form $\delta_r(B)Y_t = \omega_r(B)B^b X_t$

rsb	∇ form	B form	Impulse response v_j	
00b	$Y_t = g X_{t-b}$	$Y_t = \omega_0 B^b X_t$	0	$j<b$
			ω_0	$j=b$
			0	$j>b$
01b	$Y_t = g(1+\eta_1\nabla)X_{t-b}$	$Y_t = (\omega_0 - \omega_1 B)B^b X_t$	0	$j<b$
			ω_0	$j=b$
			$-\omega_1$	$j=b+1$
			0	$j>b+1$
02b	$Y_t = g(1+\eta_1\nabla+\eta_2\nabla^2)X_{t-b}$	$Y_t = (\omega_0 - \omega_1 B - \omega_2 B^2)B^b X_t$	0	$j<b$
			ω_0	$j=b$
			$-\omega_1$	$j=b+1$
			$-\omega_2$	$j=b+2$
			0	$j>b+2$
10b	$(1+\xi_1\nabla)Y_t = g X_{t-b}$	$(1-\delta_1 B)Y_t = \omega_0 B^b X_t$	0	$j<b$
			ω_0	$j=b$
			$\delta_1 v_{j-1}$	$j>b$
11b	$(1+\xi_1\nabla)Y_t = g(1+\eta_1\nabla)X_{t-b}$	$(1-\delta_1 B)Y_t = (\omega_0 - \omega_1 B)B^b X_t$	0	$j<b$
			ω_0	$j=b$
			$\delta_1\omega_0 - \omega_1$	$j=b+1$
			$\delta_1 v_{j-1}$	$j>b+1$
12b	$(1+\xi_1\nabla)Y_t = g(1+\eta_1\nabla+\eta_2\nabla^2)X_{t-b}$	$(1-\delta_1 B)Y_t = (\omega_0 - \omega_1 B - \omega_2 B^2)B^b X_t$	0	$j<b$
			ω_0	$j=b$
			$\delta_1\omega_0 - \omega_1$	$j=b+1$
			$\delta_1^2\omega_0 - \delta_1\omega_1 - \omega_2$	$j=b+2$
			$\delta_1 v_{j-1}$	$j>b+2$

20b $\quad (1+\xi_1\nabla+\xi_2\nabla^2)Y_t = gX_{t-b}$

$\qquad\qquad (1-\delta_1 B-\delta_2 B^2)Y_t = \omega_0 B^b X_t$

$\qquad\qquad\qquad\qquad$ $0 \qquad\qquad\qquad\qquad\qquad j<b$
$\qquad\qquad\qquad\qquad$ $\omega_0 \qquad\qquad\qquad\qquad\qquad j=b$
$\qquad\qquad\qquad\qquad$ $\delta_1 v_{j-1}+\delta_2 v_{j-2} \qquad\qquad j>b$

21b $\quad (1+\xi_1\nabla+\xi_2\nabla^2)Y_t = g(1+\eta_1\nabla)X_{t-b}$

$\qquad\qquad (1-\delta_1 B-\delta_2 B^2)Y_t = (\omega_0-\omega_1 B)B^b X_t$

$\qquad\qquad\qquad\qquad$ $0 \qquad\qquad\qquad\qquad\qquad\quad j<b$
$\qquad\qquad\qquad\qquad$ $\omega_0 \qquad\qquad\qquad\qquad\qquad\quad j=b$
$\qquad\qquad\qquad\qquad$ $\delta_1\omega_0-\omega_1 \qquad\qquad\qquad j=b+1$
$\qquad\qquad\qquad\qquad$ $\delta_1 v_{j-1}+\delta_2 v_{j-2} \qquad\quad j>b+1$

22b $\quad (1+\xi_1\nabla+\xi_2\nabla^2) = g(1+\eta_1\nabla+\eta_2\nabla^2)X_{t-b}$

$\qquad\qquad (1-\delta_1 B-\delta_2 B^2)Y_t = (\omega_0-\omega_1 B-\omega_2 B^2)B^b X_t$

$\qquad\qquad\qquad\qquad$ $0 \qquad\qquad\qquad\qquad\qquad\qquad\qquad\qquad j<b$
$\qquad\qquad\qquad\qquad$ $\omega_0 \qquad\qquad\qquad\qquad\qquad\qquad\qquad\qquad j=b$
$\qquad\qquad\qquad\qquad$ $\delta_1\omega_0-\omega_1 \qquad\qquad\qquad\qquad\qquad j=b+1$
$\qquad\qquad\qquad\qquad$ $(\delta_1^2+\delta_2)\omega_0-\delta_1\omega_1-\omega_2 \quad j=b+2$
$\qquad\qquad\qquad\qquad$ $\delta_1 v_{j-1}+\delta_2 v_{j-2} \qquad\qquad\qquad j>b+2$

$$\xi_1=\frac{\delta_1+2\delta_2}{1-\delta_1-\delta_2} \qquad \xi_2=\frac{-\delta_2}{1-\delta_1-\delta_2} \qquad\qquad \delta_1=\frac{\xi_1+2\xi_2}{1+\xi_1+\xi_2} \qquad \delta_2=\frac{-\xi_2}{1+\xi_1+\xi_2}$$

$$g=\frac{\omega_0-\omega_1-\omega_2}{1-\delta_1-\delta_2} \qquad\qquad\qquad\qquad \omega_0=\frac{g(1+\eta_1+\eta_2)}{1+\xi_1+\xi_2}$$

$$\eta_1=\frac{\omega_1+2\omega_2}{\omega_0-\omega_1-\omega_2} \qquad\qquad\qquad\qquad \omega_1=\frac{g(\eta_1+2\eta_2)}{1+\xi_1+\xi_2}$$

$$\eta_2=\frac{-\omega_2}{\omega_0-\omega_1-\omega_2} \qquad\qquad\qquad\qquad \omega_2=\frac{-g\eta_2}{1+\xi_1+\xi_2}$$

$$1-\delta_1-\delta_2=(1+\xi_1+\xi_2)^{-1}$$

Source: George E. P. Box and Gwilym M. Jenkins. Reproduced from *Time Series Analysis: Forecasting and Control*, Holden-Day, Oakland, CA 1976.

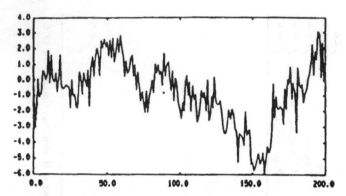

Figure 7.6 Illustrative plot of a first-order ARIMA model.

The goal here is to determine the parameters ϕ and θ of the noise model. For this purpose, we compute the autocorrelation functions and partial autocorrelation functions and refer to appropriate figures and tables to obtain estimates of the parameters. Figure 7.6 shows an illustrative plot for first-order ARIMA models. Information on other types of models may be found in the *Tables for Statisticians* [7].

Step 5. Find better estimates of the parameters. The problem can now be posed as follows. Given a set of $n = N - d$ values of the deviation variables Y_t and X_t (y_t and x_t if $d > 0$) along with the initial estimates b, δ, ω, ϕ, θ, and the starting values x_0, y_0, and a_0, find the maximum likelihood estimates of the parameters of the transfer function and the noise model which minimize the conditional sum of squares function given by the expression

$$S_0(b, \delta, \omega, \phi, \theta) = \sum_{t=1}^{n} a_t^2 (b, \delta, \omega, \phi, \theta \mid x_0, y_0, a_0) \qquad (7.101)$$

where the starting values x_0, y_0, a_0, represent the values of these variables prior to the commencement of the series. Given a set of starting values and parameters, the a's can be computed as follows. Recall that the combined model is

$$y_t = \delta^{-1}(B)\omega(B)x_{t-b} + n_t \qquad (7.102)$$

with

$$n_t = \phi^{-1}(B)\theta(B)a_t \qquad (7.103)$$

First, compute the contribution of the transfer function model to the output according to

$$\hat{y}_t - \delta_1\hat{y}_{t-1} - \cdots - \delta_r\hat{y}_{t-r} = \omega_0 x_{t-b} - \omega_1 x_{t-b-1} - \cdots - \omega_s x_{t-b-s}$$

$$(7.104)$$

Then the noise series can be computed from

$$n_t = y_t - \hat{y}_t \tag{7.105}$$

Finally the a's can be computed according to

$$a_t = \theta_1 a_{t-1} + \cdots + \theta_q a_{t-q} + n_t - \phi_1 n_{t-1} - \cdots - \phi_p n_{t-p} \tag{7.106}$$

For stochastic estimation, the effect of transients can be minimized by starting at $t = u + 1$, where u is the larger of r and $s + b$. Thus, the conditional sum of squares function takes on the expression

$$S_0(b, \delta, \omega, \phi, \theta) = \sum_{t=u+p+1}^{n} a_t^2 (b, \delta, \omega, \phi, \theta, |x_0, y_0, a_0) \tag{7.107}$$

A nonlinear least squares algorithm has been suggested for the purpose of this minimization. Computer programs with this subroutine are available on many main frame machines.

Step 6. Carry out diagnostic checks and modify models as necessary. This is the final stage of the stepwise procedure.

Suppose that the combined transfer function–noise model is

$$y_t = \delta^{-1}(B)\omega(B)x_{t-b} + \phi^{-1}(B)\theta(B)a_t = h(B)x_t + \psi(B)a_t \tag{7.108}$$

but a wrong model is used in the time series analysis given by

$$y_t = h_0(B)x_t + \psi_0(B)\hat{a}_t \tag{7.109}$$

Equations (7.108) and (7.109) may be solved for the residuals $\hat{a}_t$, giving

$$\hat{a}_t = \psi_0(B)^{-1}\{h(B) - h_0(B)\}x_t + \psi_0^{-1}(B)\psi(B)a_t \tag{7.110}$$

The estimated autocorrelation function $r_{\hat{a}\hat{a}}(k)$ and the cross-correlation function $r_{x\hat{a}}(k)$ are computed leading to the following conclusions

a. If the residuals are autocorrelated but are not cross-correlated with the x_t's, then the transfer function model is correct but the noise model is not. The form of $r_{\hat{a}\hat{a}}(k)$ may indicate appropriate changes which should be made to the noise model.

b. If the residuals are autocorrelated *and* are cross-correlated with the x_t's, then the transfer function is correct but the noise model may or may not be correct. The cross-correlation function could suggest the modifications to the transfer function model.

As a final over check a χ^2 test on the quantity

$$Q = m \sum_{k=1}^{K} r_{\hat{a}\hat{a}}^2(k) \tag{7.111}$$

and

$$S = m \sum_{k=0}^{K} r_{a\hat{a}}(k) \tag{7.112}$$

has been suggested. Here α is the input in prewhitened form and K is sufficiently large such that for $j > K$ the values of $\psi_j(B)$ from the model

$$y_t = h(B)x_t + \psi(B)a_t \tag{7.113}$$

are negligible. With Q and S and a table of percentage points of χ^2, one can obtain an approximate test of hypothesis of model adequacy.

7.5.3 Time Series Analysis for Multivariable Systems

The foregoing procedure may be extended to multivariable systems. Initial estimates of impulse response coefficients may be obtained by expanding the cross-correlation and autocorrelation matrices of the previous section for each output. The initial estimates may be fine tuned by an optimization procedure along the lines discussed in the previous subsection.

7.6 A Random Search Procedure for Identification

The starting point for the random search optimization procedure [8] is a time domain equation relating an output to an input that contain the parameters to be identified. The identification problem is, given a set of experimental input–output data, to find the values of the parameters of an assumed process model such that the model-predicted data fit the experimental data well. The procedure is simple and can be used with equal ease for single as well as multiparameter identification, but the execution times increase dramatically as the number of parameters increases. We illustrate the procedure with reference to an overdamped second-order plus deadtime process model where the goal is to determine the parameters K_p, θ_d, τ_1, and τ_2. In this instance, the time domain equation relating an output Y to an input X is given by

$$Y_i = a_1 Y_{i-1} - a_2 Y_{i-2} + b_1 X_{i-N-1} + b_2 X_{i-N-2} + b_3 X_{i-N-3} \tag{7.114}$$

The procedure consists of the following steps:

1. Introduce a suitable disturbance (e.g., step or pulse) at the input of the process while the process is operating at steady state in manual control.
2. Record the input and output data.
3. Assume trial values of K_P, θ_d, τ_1, and τ_2.
4. Using these trial values and the input data of Step 1, predict the output data via Eq. (7.114).
5. Calculate the difference between the experimental output and the predicted output, an indication of error in the trial parameters, by the expression.

$$\text{ERROR} = \sum_{i=0}^{N} (Y_{ia} - Y_{ip})^2 \qquad (7.115)$$

6. Update the parameters and repeat Steps 4 and 5 until the ERROR is minimized to an acceptable level.

From these six steps it should be clear that this is a classical optimization problem having Eq. (7.115) as the objective function. These steps are embedded in the random search procedure as follows:

1. Specify initial values of the parameters X_1, X_2, X_3, X_4 to be optimized and denote them as X_1^0, X_2^0, X_3^0, X_4^0 and specify an initial range for each r_1^0, r_2^0, r_3^0, r_4^0. Set iteration index $j = 1$.
2. Read in a sufficient number of random numbers between -0.5 and $+0.5$ (2000 numbers are suggested by Luus and Jaakola [8]).
3. Take four P random numbers, $R(K,i)$, from Step 2 (Luus and Jaakola suggest $P = 100$) and assign them to X_1, X_2, X_3, X_4 so that there are P sets of values, which are calculated according to

$$X_i^j = X_i^0 + R(K,i)r_i^{j-1} \qquad (7.116)$$

where

$i = 1, ..., 4$
$k = 1, ..., P$

4. Test constraint equations and calculate a value of ERROR for each set. In this instance, the constraints are that K_P and $\theta_d > 0$. Also the definitions of the constants a_1, a_2, b_1, b_2, b_3 are such that to avoid numerical problems on the computer it is necessary to specify that

$$\tau_1 - \tau_2 < \epsilon_1 \qquad (7.117)$$

$$\tau_1 > \epsilon_1 \qquad (7.118)$$

$$\tau_2 > \epsilon_2 \qquad (7.119)$$

5. From the P sets, find the set having minimum ERROR. Write this ERROR and the corresponding values of X_i and denote them as X_i^0.
6. If the number of iterations has reached the maximum (Luus and Jaakola suggest 200 iterations), then end the problem. If not, go to the next step.
7. Reduce the range by an amount ϵ such that

$$r_i^j = (1 - \epsilon)r_i^{j-1} \tag{7.120}$$

Usually, $\epsilon = 0.05$.
8. Increase j by 1 to $j + 1$ and then go to Step 3 and continue.

A flowchart of the program for the implementation of the random search procedure is shown in Fig. 7.7. This procedure has been successfully used to find 15 kinetic constants of an industrial polymerization model such that the model output data on monomer conversion, etc., adequately represented the experimental data. For proprietary reasons, additional details cannot be revealed at this time.

7.7 Applications to Polymerization Reactors

In the subsequent chapters, applications of the relevant control technologies to polymerization reactors will be included. Efforts will be made to indicate the unique ways in which the general techniques have been tailored to the specific polymerization situation. In addition to the general features of polymerization control (poor sensors and models, strong nonlinearities, etc.), each polymerization reactor has its own specific requirements. It is the belief of the authors (and one of the themes of this book) that good control of polymerization reactors can only be achieved when control theory is combined with an understanding of the principles of polymerization reactions *and* the specific requirements of the current application.

7.7.1 Transfer Function Modeling of a Continuous Emulsion Polymerization Reactor

Leffew and Deshpande [9] studied the control of a continuous (CSTR) emulsion polymerization of vinyl acetate. These were simulation studies in which the "plant" was a detailed model of the emulsion polymerization of vinyl acetate developed by Kiparissides et al. [10]. Monomer conversion was controlled by manipulating either initiator flow rate or reactor temperature. Identification was carried out by making a step change in the manipulated variable and observing the response of the detailed mathematical model to this input. A second-order plus deadtime model was assumed, and the method of Brantley [11] was used to determine the model parameters. The results are shown in Table 7.2. It is interesting to note that a very complex system, approximated by a detailed model [9] could be approximated by a rather simple second-order plus deadtime model.

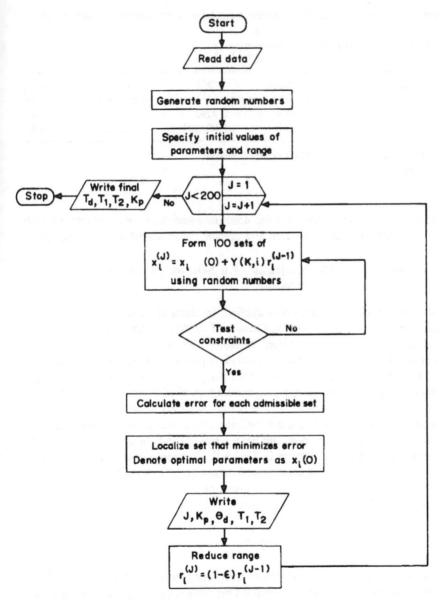

Figure 7.7 Random search procedure flowchart.

Table 7.2 Identification of a continuous emulsion polymerization via step testing.

Model form: $G_p(s) = \dfrac{K_p e^{-\theta_d s}}{(\tau_1 s - 1)(\tau_2 s + 1)}$

Manipulated variable	θ_d (min)	$K_p{}^a$	τ_1 (min)	τ_2 (min)
Initiator flow	30.0	0.2793	3.6826	0.0200
Reactor temperature	30.0	0.002529	2.2540	0.0200

^{a}lb/hr/fractional conversion of °C/fractional conversion.
Source: Ref. 9.

It is also interesting to note that the deadtimes and the time constants for both manipulations (initiator flow rate and reactor temperature) are quite similar.

7.7.2 Step Response Modeling of a Semibatch Solution Polymerization Reactor

A nonlinear DMC algorithm was applied, in simulation, to the control of the semibatch reactor for the polymerization of methyl methacrylate described in Subsection 3.4.3. The manipulated variables were initiator flow, u_1, and cooling jacket temperature, u_2. The controlled outputs are the reaction temperature, y_1, and the number average molecular weight, y_2. This model poses some unique problems and challenges for applying nonlinear predictive control. First, the batch nature of the process means there is really no steady state.

For temperature control, the DMC algorithm uses a step response model. Because the batch reactor does not have a true "steady state," the step coefficients are determined as shown in Fig. 7.8. Keeping a constant input and integrating forward gives the unperturbed output y_{past}. Stepping the input to a new value and integrating into the future gives the perturbed output, y_{per}. The step response coefficients are then calculated from

$$a_i = \frac{y_{per}(k + i) - y_{past}(k + i)}{\Delta u} \tag{7.121}$$

In the case of molecular weight control, open-loop simulation revealed that the initiator had very different effects on the molecular weight depending on the time it was introduced to the batch reactor [13]. This was best observed by adding an initiator *pulse* of 12-min duration (i.e., one sampling time) at different times during the reaction rather than introducing a step input which lasts over the entire time of reaction. Consequently, instead of the step model, an impulse model of the following form [14] is used in the DMC algorithm for the molecular weight:

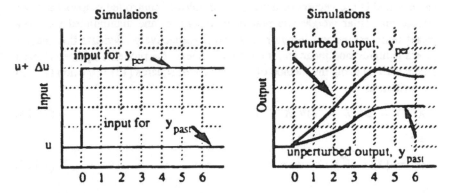

Figure 7.8 Step response coefficients for a semibatch polymerization reactor. Estimating step coefficients from an unsteady-state nonlinear model. Two simulations are run at different values for input. For unperturbed simulation, input $= u$; for the perturbed simulation, input $= u + \Delta u$. The ith step coefficient can be approximated as $a_i = (y_{per,i} - y_{past,i})/\Delta u$. (From Ref. 13.)

$$y(k) = \sum_{i=1}^{N} h_i u(k - i) \qquad (7.122)$$

The impulse coefficients, h_i, are calculated similar to the step response coefficients by integrating the nonlinear system with a pulse input instead of a step. The use of these system models in the nonlinear DMC application will be discussed in Chapter 10.

7.7.3 Time Series Identification of a Semibatch Solution Polymerization Reactor

An analytical model of polymerization reactions is very important in improving the control of the system; however, the kinetics of most polymerization reactions are not accurately known. To deal with this, an empirical model of the process dynamics can be employed. In this simulation study [12], the semibatch solution polymerization of methyl methacrylate was controlled using a short-horizon adaptive–predictive controller. This controller adjusts the manipulated variable(s) to bring the output(s) to a desired value over a short horizon. The simulation model was detailed in Subsection 3.4.3. Although, as we have seen, the model is highly nonlinear, a linear time series model (control model) was used for the control calculations. The discrete control model was of the following form:

$$X(K + 1) = \phi(K)X(K) + \beta(K)U(K) \qquad (7.123)$$

where X and U are vectors of deviation variables of controlled outputs (number average molecular weight and reactor temperature) and manipulated variables (jacket temperature and initiator flow rate), respectively. Because this reactor system is highly nonlinear and the model is linear, the model is updated at each sample interval using the method of recursive least squares [15]. Equation (7.123) is rewritten as

$$X(K) = \phi^T(K - 1)\theta(K - 1) \tag{7.124}$$

where

$$\phi^T = [X^T \vdots U^T] \tag{7.125}$$

and

$$\Theta = \begin{bmatrix} \phi^T \\ \cdots \\ \beta^T \end{bmatrix} \tag{7.126}$$

The parameter vectors, Θ_i (the columns of Θ), are updated as follows:

$$P_i(K|K - 1) = P_i(K - 1) + D_i \tag{7.127}$$

$$P_i(K) = P_i(K|K - 1) - P_i(K|K - 1)\phi(K - 1)$$
$$\times \ [\phi^T(K - 1)P_i(K|K - 1)\phi(K - 1) + 1]^{-1}$$
$$\phi^T(K - 1)P_i(K|K - 1) \tag{7.128}$$

$$\theta_i(K) = \theta_i(K - 1) + P_i(K)\phi(K - 1)[X_i(K) - \phi^T(K - 1)\theta_i(K - 1)] \tag{7.129}$$

Equations (7.127)–(7.129) are implemented for each reactor state.

One major advantage of adaptive–predictive control is its ability to adapt to changing process conditions. To demonstrate this advantage, a process distur- bance was applied to both the baseline and adaptive–predictive control cases with all of the control parameters remaining the same. The disturbance selected was reduction in the initiator efficiency by 20%. This may be thought to represent the normal lot-to-lot variation in monomer and initiator reactivity. When this distur- bance is applied to the baseline case, the batch time is increased by 10% and the molecular weight is increased by 9%. When the same disturbance is applied to the adaptive–predictive controller, more initiator is added to the reactor to com- pensate for the reduced initiator efficiency. After an initial period of adjustment, the molecular weight remains on target throughout the batch cycle. Figure 7.9 shows the adaptation of one of the time series model parameters (the effect of initiator addition on monomer conversion), as it first converges from an initial

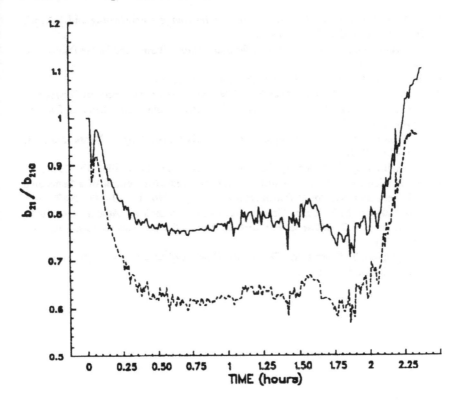

Figure 7.9 Model coefficient adaptation. Adaptive model coefficient b_{21} (effect of initiator addition on monomer conversion) adaptation. b_{21} is normalized by its initial value, b_{210} (—normal initiation; --- poor quality initiator.) (From Ref. 12.)

estimate, then adapts in response to the nonlinearity induced by the gel effect. Note that whereas the trend is the same, the values are offset for the case of a poor quality initiator.

References

1. W. H. Ray, *Advanced Process Control*, McGraw-Hill, New York, 1981.
2. C. L. Smith, *Digital Computer Process Control*, Intext Publishers, Scranton, PA, 1972.
3. W. C. Clements and K. B. Schnelle, *Ind. Eng. Chem. Proc. Des. Dev.*, 2, 94 (1963).
4. J. O. Hougen, *Experiences and Experiments with Process Dynamics*, CEP Monograph Series 60, Am. Inst. of Chem. Eng., New York, 1964, p. 4.
5. W. L. Luyben, *Process Modeling, Simulation, and Control for Chemical Engineers*, McGraw-Hill, New York, 1973.

6. G. E. P. Box and G. M. Jenkins, *Time Series Analysis Forecasting and Control*, Holden-Day, Oakland, CA, 1976.
7. H. Arkins and R. R. Colton, *Tables for Statisticians*, Barnes and Noble Publishers, New York, 1963
8. R. Luus and T. H. Jaakola, AIChE J., 19, 760 (1973).
9. K. W. Leffew and P. B. Deshpande, in *Emulsion Polymers and Emulsion Polymerization* (D. R. Bassett and A. E. Hamielec, eds.), American Chemical Society, Washington, D.C., 1981.
10. C. Kiparissides, J. F. McGregor, and A. E. Hamielec, *J. Appl. Polym. Sci.*, 23, 401–418 (1979).
11. R. O. Brantley, M.S. thesis, University of Louisville, Louisville, KY, 1981.
12. W. E. Houston and F. J. Schork, "Adaptive Predictive Control of a Semibatch Polymerization Reactor," *Polym. Process Eng.*, 5, No. 1, 119–147 (1987).
13. Tod Peterson, M.S. thesis, Georgia Institute of Technology, Atlanta, GA, 1990.
14. D. M. Prett and C. E. García, *Fundamental Process Control*, Butterworths, Needham, MA, 1988.
15. L. Ljung and T. Soderstrom, *Theory and Practice of Recursive Identification*, MIT Press, Cambridge, MA, 1983.

8

Single-Loop Reactor Control Strategies

As we have seen in Chapter 5, the complete description of a polymerization reactor control problem involves multiple inputs and outputs. Typical outputs are reactor temperature, monomer conversion, and average molecular weight/molecular weight distribution. The manipulated variables available are the flow rates of monomer(s) and initiator, rate of agitation (in some cases), and cooling (for exothermic reactions). Although the problem is multivariable in nature, a SISO control strategy is employed in many industrial applications owing to the difficulties mentioned earlier. In these instances, reactor temperature is the primary variable that is controlled, with the others being monitored/controlled in an open-loop manner. PID type algorithms have been used in these applications for the most part. With the advances in modeling and sensing instrumentation, it has become possible in some applications to control some of the other outputs as well. If multiloop control structure is selected in such cases, then single-input single-output (SISO) control algorithms would be utilized as control elements for each of the multiple loops. In this chapter, we take up the design of single-input single-output algorithms and their application to polymerization reactors.

Temperature control in exothermic polymerization reactors is achieved by manipulating the flow rate of a cooling medium flowing through the jacket or by bypassing a portion of chilled monomer feed (in some cases) to the reactor. Design of temperature controllers for polymerization reactors warrants special consideration because the exothermicity of reactions leads to potentially open-loop unstable behavior. Temperature control strategies for batch and continuous reactors have slightly different aims. In batch systems, the goal may be to bring the reactor contents from ambient conditions to the reactor temperature setpoint in a smooth manner and then to operate the reactor at the setpoint temperature for the remaining duration of the batch. The first phase involves heating, whereas the second phase involves cooling. In some cases, the second phase may involve a desired temperature profile which must be maintained rather than a constant setpoint. Also, if a deviation in the setpoint profile were to occur due to load disturbances, it may not be sufficient to simply bring the reactor outputs back to the original profile. A new optimization problem may have to be solved to determine the future setpoint profile that the reactor output(s) must follow if the end products are to meet the produce specifications. Furthermore, the time at which the switch should be made from heating to cooling is a relevant issue in batch reactor temperature control. In continuous reactors, there is a start-up period as in batch systems, but once the desired operating conditions are reached, the reactor is operated at these conditions indefinitely. In continuous systems, a number of polymer grades are often produced. These grades may involve monomers that are common to all of them but specialty chemical(s) may be added that are unique to each grade. How to achieve a grade switch without producing an excessive amount of off-spec material is a relevant issue here.

SISO controllers may be designed by a number of methods. Among them are frequency domain methods, pole placements methods, and the response specification method. Since the advent of digital computers for process control, a number of control algorithms have been developed by the response specification method. We shall adopt this method in this chapter.

It is possible to design controllers for setpoint changes or for load changes. Due to the availability of computers for process control, a number of digital control algorithms have been developed during the last three decades. Several of these algorithms are derived in the following sections. The treatment is brief and the interested reader may refer to Ref. 1 for additional details.

8.1 Servo Design

The block diagram of a typical sampled data control structure is shown in Fig. 8.1. The closed-loop pulse transfer function relating the controlled variable to the setpoint is

$$\frac{C}{R} = \frac{G_c G}{1 + G_c G} \tag{8.1}$$

where

R = setpoint,
G_c = controller transfer function, $M(z)/E(z)$,

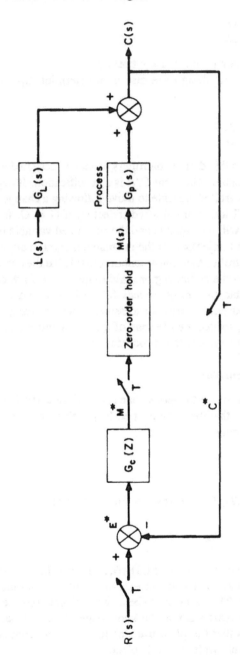

Figure 8.1 Typical sampled data control system (stirred functions indicate discrete signals).

$G = z\{G_{h0}(s)\ G_p(s)\}$,
C = controlled variable.

The z-transform operator is omitted for brevity.

Because the goal of the design exercise is to determine G_C, Eq. (8.1) may be rearranged to give

$$G_c = \frac{C/R}{1 - C/R}\ \frac{1}{\tilde{G}} \tag{8.2}$$

Equation (8.2) shows that digital controllers cannot be designed unless a model of the plant, $\tilde{G}$, is available. The term C/R is a specification. It depicts how the designer wishes the controlled variable to move following a change in point. The controlled variable will follow the desired trajectory if $G = \tilde{G}$. In the presence of modeling errors, G will not equal $\tilde{G}$ and the controlled variable trajectory will deviate from the desired trajectory. If the mismatch is significant, even instability may occur. The controller then must contain suitable handles which will bring the operation back into the stable region. Although Eq. (8.2) does not contain the manipulated variable explicitly, successful designs would ensure that the manipulated variable movements are not excessive. Excessive manipulated variable movements, of course, reduce the life span of the final control element and pose a burden on the control system that processes them.

8.1.1 Deadbeat Control

Deadbeat control gives the best response one can achieve under ideal conditions. For this case, assuming that the plant is open-loop stable, the specified response is written down according to

$$\frac{C}{R} = z^{-(N+1)} \tag{8.3}$$

Substituting in Eq. (8.2), this expression for C/R gives

$$G_c = \frac{z^{-(N+1)}}{1 - z^{-(N+1)}}\ \frac{1}{\tilde{G}} \tag{8.4}$$

The specification depicted in Eq. (8.3) requires that the controlled variable shall reach the setpoint as soon as the effects of sampling deadtime and the process deadtime are over. This turns out to be a very stringent requirement in practical applications. One reason is that the knowledge of plant dynamics is rarely precise. The second is that the plant transfer function may contain zeroes that are too close to the unit circle in the Z plane.

Now, suppose that a deadbeat control algorithm is to derived for a first-order plant containing N sampling instants of deadtime. For this case,

$$\check{G} = z\{G_{ho}G_p(s)\}$$

$$= z\left\{\frac{1 - e^{-sT}}{s} \frac{K_p e^{-\theta ds}}{\tau_p S + 1}\right\}$$

$$= \frac{K_p(1 - \alpha)z^{-(N+1)}}{1 - \alpha z^{-1}} \tag{8.5}$$

where, $\alpha = e^{-T/\tau_p}$. Substituting for $\check{G}$ into Eq. (8.4) gives

$$G_c(z) = \frac{M(z)}{E(z)} = \frac{z^{-(N+1)}}{1 - z^{-(N+1)}} \frac{1 - \alpha z^{-1}}{K_p(1 - \alpha)z^{-(N+1)}}$$

$$= \frac{1}{K_p(1 - \alpha)} \left\{\frac{1 - \alpha z^{-1}}{1 - z^{-(N+1)}}\right\} \tag{8.6}$$

Cross-multiplying terms in Eq. (8.6), followed by inversion, gives

$$M_n = M_{n-(N+1)} + \frac{1}{K_p(1 - \alpha)} [E_n - \alpha E_{n-1}] \tag{8.7}$$

The application of Eq. (8.7) will result in deadbeat control in the absence of modeling errors.

8.1.2 Dahlin Algorithm

In this approach [2] it is specified that the closed-loop respond as though it were a first-order with deadtime process. Thus, for this case,

$$\frac{C}{R} = \frac{(1 - \beta)}{1 - \beta z^{-1}} z^{-(N+1)} \tag{8.8}$$

where $\beta = e^{-\tau/\tau_{cl}}$. Substituting this specification in Eq. (8.2) gives

$$G_c = \frac{(1 - \beta)z^{-(N+1)}}{1 - \beta z^{-1} - (1 - \beta)\check{z}^{-(N+1)}} \frac{1}{\check{G}} \tag{8.9}$$

If the process is first order with deadtime having the transfer function given in Eq. (8.5), then Eq. (8.9) becomes

$$G_c(z) = \frac{M(z)}{E(z)}$$

$$= \frac{(1 - \beta)z^{-(N+1)}}{1 - \beta z^{-1} - (1 - \beta)z^{-(N+1)}} \frac{1 - \alpha z^{-1}}{K_p(1 - \alpha)z^{-(N+1)}} \tag{8.10}$$

or

$$G_c(z) = \frac{1 - \beta}{K_p(1 - \alpha)} \left[\frac{1 - \alpha z^{-1}}{1 - \beta z^{-1} - (1 - \beta)z^{-(N+1)}} \right]$$

Again, the time domain version of the Dahlin control algorithm may be determined by cross-multiplying terms in Eq. (8.10) and inverting the result giving

$$M_n = \beta M_{n-1} + (1 - \beta)M_{n-(N+1)} + \frac{1 - \beta}{K_p(1 - \alpha)}[E_n - \alpha E_{n-1}] \tag{8.11}$$

The term β $(0 < \beta < 1)$ determines the speed of response. Smaller values of β demand faster responses, but the system becomes very sensitive to modeling errors, and vice versa. In practical applications, a suitable of β of must be found that represents a good compromise between robustness and dynamic performance.

8.1.3 Internal Model Control

In this approach (IMC) [3], the plant model is factorized according to

$$\hat{G} = G_+ G_- \tag{8.12}$$

where

G_+ = nonminimum phase elements (deadtime, inverse response),
G_- = remaining part of G.

The specification for C/R is written down as

$$C/R = FG_+ \tag{8.13}$$

where the term G_+ accommodates the nonminimum phase elements and F is a filter which can be used for loop shaping and to ensure robustness in the presence of modeling errors. Substituting the above expression for C/R into Eq. (8.2) gives

$$G_c = \frac{FG_I}{1 - FG_+} \tag{8.14}$$

where $G_I = 1/G_-$.
  Now, from Eq. (8.12)

$$G_+ = \frac{\hat{G}}{G_-} = \hat{G}G_I \tag{8.15}$$

Substituting for G_+ in Eq. (8.14) gives

$$G_c = \frac{FG_I}{1 - FG_I\hat{G}} \tag{8.16}$$

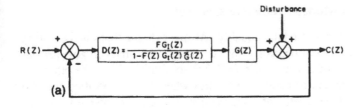

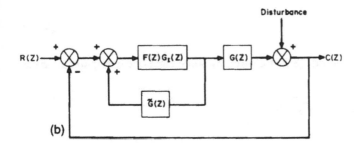

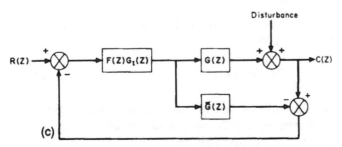

Figure 8.2 Derivation of IMC structure by response specification method. (a) Basic block diagram; (b) equivalent block diagram; (c) IMC structure.

The control law given in Eq. (8.16) is shown in Fig. 8.2a or equivalently in Figs. 8.2b and 8.2c. Figure 8.2c depicts the IMC structure, but the three figures are entirely equivalent.

The term $F(z)$ is selected such that the roots of the characteristic equation $1 + DG = 0$ can be brought back inside the unit circle in the z-domain in the presence of modeling errors. If a first-order transfer function is selected to represent F, then IMC and the Dahlin algorithm are the same.

8.1.4 Simplified Model-Predictive Control

There are two spectrums of control quality. At one extreme there is the notion of perfect control (deadbeat control). IMC is an algorithm that can give perfect control in the absence of modeling errors if F is set equal to 1. At the other extreme, there is the notion of open-loop control wherein it is specified that the control algorithm deliver a setpoint response that is at least as good as the open-loop response. Simplified model-predictive control [4] (SMPC) utilizes this type of specification for design purposes. Thus,

$$\frac{C}{R} = \frac{1}{K_p} \hat{G} \tag{8.17}$$

substituting for C/R in Eq. (8.2) followed by some algebraic manipulation gives

$$G_c(z) = \frac{M(z)}{E(z)} = \frac{1}{K_p - \hat{G}} \tag{8.18}$$

or

$$M(z) = \frac{1}{K_p} E(z) + \frac{1}{K_p} \hat{G}(z)M(z) \tag{8.19}$$

Now, the closed-loop response may be speeded up by replacing the term $1/K_p$ by a tuning constant, say $\bar{\alpha}$, giving

$$M(z) = \bar{\alpha}E(z) + \frac{1}{K_p} \hat{G}(z)M(z) \tag{8.20}$$

At this point, a z-domain transfer function model may be substituted for $\hat{G}$ and the equation may be inverted into the time domain or alternatively an impulse response model may be employed, giving

$$M(z) = \bar{\alpha}E(z) + \frac{1}{K_p} \sum_{i=1}^{p} h_i z^{-i} M(z) \tag{8.21}$$

Inversion of Eq. (8.21) gives the SMPC control law in the time domain as

$$M_n = \bar{\alpha}E_n + \frac{1}{K_p} \sum_{i=1}^{p} h_i M_{n-i} \tag{8.22}$$

The tuning constant $\bar{\alpha}$ determines the speed of response. In the absence of modeling errors, the loop may be tightly tuned. In the presence of modeling errors, $\bar{\alpha}$ may be adjusted to ensure robustness. The impulse response coefficients may be obtained from the experimental step response coefficients according to the procedure previously described. The use of step response coefficients circumvents the problem of fitting experimental data to parametric models; often experimental data cannot be adequately fitted to parametric models.

8.1.5 Conservative Model-Based Control

The SMPC algorithm may be extended to provide deadtime compensation by replacing the term $1/K_p$ in Eq. (8.19) in both places by $(1 - \bar{\beta}z^{-1})/K_p(1 - \bar{\beta})$, giving

$$M(z) = \frac{(1 - \bar{\beta}z^{-1})}{K_p(1 - \bar{\beta})} [E(z) + \check{G}(z)M(z)] \qquad (8.23)$$

or

$$M_n = \frac{1}{K_p(1 - \bar{\beta})} \left[E_n - \bar{\beta}E_{n-1} + \sum_{i=1}^{p} h_i M_n - \bar{\beta} \sum_{i=1}^{p} h_i M_{n-i-1} \right]$$

where $\bar{\beta}$ ($0 < \bar{\beta} < 1$) is the tuning constant of the algorithm. This is known as conservative model-based control (CMBC) [5].

8.1.6 Smith Predictor and Analytical Predictor

The Smith predictor [6] and analytical predictor [7] contain a PID type controller but they contain additional blocks to provide deadtime compensation. In the following, we give the classical explanation of how these algorithms are derived in the Laplace domain.

Consider a process containing deadtime given by the transfer function

$$\check{G} = G_-G_+ = G_-e^{-\theta_d s} \qquad (8.24)$$

The closed-loop transfer function for servo control is

$$\frac{C}{R} = \frac{G_c G_p}{1 + G_c G_p} = \frac{G_c G_-e^{-\theta_d s}}{1 + G_c G_-e^{-\theta_d s}} \qquad (8.25)$$

Here, G_c represents a PID type controller. The denominator of the right-hand side in Eq. (8.25) is the characteristic equation of the system. The presence of deadtime forces the designer to reduce the controller gain in G_c. Even then, poor responses can result for large values of θ_d. Equation (8.25) can be derived from Fig. 8.3a by the usual procedure.

Now, the purpose of deadtime compensation is to eliminate the delay term from the characteristic equation, giving

$$\frac{C}{R} = \frac{G_c G_-}{1 + G_c G_-} e^{-\theta_d s} \qquad (8.26)$$

Equation (8.26) is the equation that would result from the block diagram arrangement shown in Fig. 8.3b. Unfortunately, the signal B in Fig. 8.3b is fictitious. This problem may be overcome by feeding the manipulated variable signal to a mathematical model of the plant and tapping the model signal B_M for transmission to the feedback controller. The signal B and B_M will be identical if the model is the exact description of the plant as shown in Fig. 8.3c. The signal E_M representing the modeling errors and load disturbances may also be fed back as shown in Fig. 8.3d to improve performance. Figure 8.3d depicts the Smith predictor strategy. It can be easily shown that Fig. 8.3d would lead to the desired closed-loop servo transfer function given in Eq. (8.26) in the absence of modeling errors.

The analytical predictor strategy is shown in Fig. 8.4. To see the equivalence of the Smith predictor and analytical predictor strategies, Eq. (8.26) may be rearranged giving

$$\frac{Ce^{\theta ds}}{R} = \frac{G_c G_-}{1 + G_c G_-} \tag{8.26a}$$

The block diagram shown in Fig. 8.4 has the transfer function given in Eq. (8.26a). Note again that deadtime has been eliminated from the characteristic

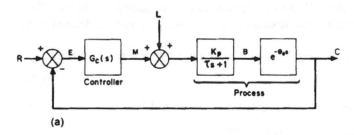

(a)

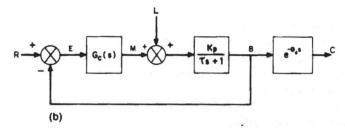

(b)

Figure 8.3 Derivation of the Smith predictor. (a) Conventional feedback loop containing deadtime; (b) desired configuration of the feedback loop; (c) preliminary Smith predictor scheme; (d) final Smith predictor scheme.

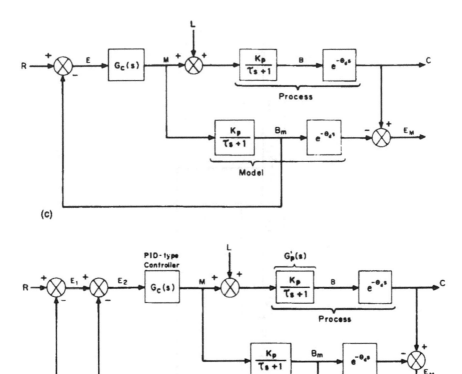

(c)

(d)

Figure 8.3 Continued

equation of the system. The difficulty posed by the presence of the unreliable element $e^{\theta_d s}$ is overcome by predicting $C(t + \theta_d)$ by the mathematical model of the process.

8.2 Regulator Design

Servo designs are useful in batch systems, start-up problems, and grade switching operations. In continuous polymerization, regulatory control is even more important. In this section, we present a method for regulatory control of

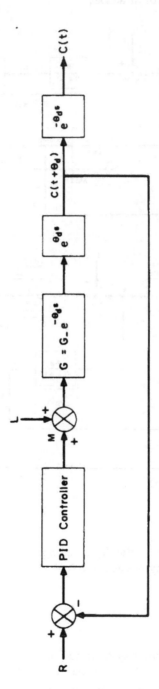

Figure 8.4 Analytical predictor strategy.

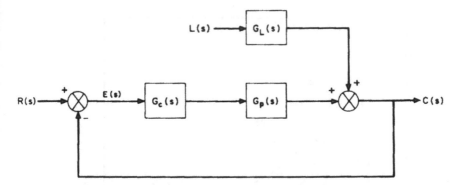

Figure 8.5 Block diagram of conventional SISO control system.

continuous systems [8]. Referring to the block diagram of an SISO conventional system shown in Fig. 8.5, the closed-loop transfer function of the system for changes in the load variable is given by

$$\frac{C}{L} = \frac{G_L}{1 + G_c G_p} \tag{8.27}$$

Again, the Laplace transform operator has been omitted for brevity. Solving Eq. (8.27) for G_c gives

$$G_c = \frac{G_L - C/L}{(C/L)G_p} \tag{8.28}$$

If $G_p = G_L$, then Eq. (8.28) reduces to

$$G_c = \frac{1}{C/L} - \frac{1}{G_p} \tag{8.29}$$

Equation (8.29) shows that for regulatory control C/L needs to be specified. Two types of specifications are envisioned. They are

$$\frac{C}{L} = \frac{s}{\tau_{cl}s + 1} e^{-\theta Ls} \tag{8.30a}$$

or

$$\frac{C}{L} = \frac{2\zeta \tau s}{\tau^2 s^2 + 2\zeta \tau s + 1} e^{-\theta Ls} \tag{8.30b}$$

Equation (8.30) expresses the desire that following a change in the load variable the controlled variable return to the setpoint via an exponential trajectory or a

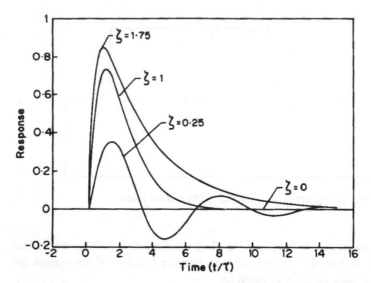

Figure 8.6 Step response to second-order trajectory specifications.

second-order trajectory. The familiar quarter decay requirements imply second-order trajectories. Illustrative plots of the second-order trajectories not containing dead time for several values ζ of τ are shown in Fig. 8.6.

For the case where $G_p = G_L$ and where the process dynamics are described by first order with deadtime models, Eqs. (8.29) and (8.30b) lead to

$$G_c(s) = \left[\left(1 - \frac{1}{K_p}\right) + \frac{1}{2\zeta\tau s} + \left(\frac{\tau}{2\zeta} - \frac{\tau_p}{K_p}\right)s\right]e^{\theta_d s} \qquad (8.31)$$

or in the time domain

$$M(t) = \left(1 - \frac{1}{K_p}\right)E(t + \theta_d) + \frac{1}{2\zeta\tau}\int_0^t E(t + \theta_d)dt$$

$$+ \left(\frac{\tau}{2\zeta} - \frac{\tau_p}{K_p}\right)\frac{dE(t + \theta_d)}{dt} \qquad (8.32)$$

Equation (8.32) is the familiar PID controller with deadtime compensation. Analytical predictor type deadtime compensation is implied in Eq. (8.32), but the Smith predictor may be used instead if desired. The unique features of Eq. (8.32) are as follows:

1. By judicious selection of the tuning constants ζ and τ, one can achieve perfect control at a high cost in terms of the manipulated variable movements or lose control at a low cost. The choice belongs to the designer. Knowing the product specifications will allow the designer to make the proper choice of ζ and τ. These are also the robustness tools available to the designer.
2. The controller constants are split into two parts. One relates to the model, whereas the other relates to the specifications. Equation (8.32) again shows the importance of good modeling.
3. Equation (8.32) show how to improve quality or reduce cost and vice versa by adjusting the tuning constants for the controller. If $G_p \neq G_L$, then the following expression results for the controller:

$$G_c(s) = \frac{M(s)}{E(s)} = \frac{\tau_P s + 1}{\tau_L s + 1} \left[\left(\frac{K_L}{K_P} - \frac{1}{K_P} \right) + \frac{K_L}{K_P 2 \zeta \tau s} + \left(\frac{\tau K_L}{K_P 2 \zeta} - \frac{\tau_L}{K_P} \right) s \right] e^{\theta_d s}$$

(8.33)

Here it is assumed that $\theta_L > \theta_d$. This equation requires a lead-lag network, a PID controller, and an analytical predictor (or Smith predictor) for implementation. These are standard blocks on modern distributed (computer) control systems. We have coined the acronym Q-PID for this type of controller. In those applications where K_L is significantly different from unity, a factor K should be inserted into the numerator of Eq. (8.30b). The appropriate sign should be selected for K giving the desired controller action (direct vs reverse).

8.3 Statistical Process Control

The recent trend toward total quality management (TQM) has generated a great deal of interest in statistical process control (SPC). This technique has been applied with good success in the discrete parts manufacturing industries. In the chemical process industries, experience has been mixed, partly because of misunderstandings about its applicability to chemical processes and its interrelation to conventional process control. We introduce SPC here because it is a technique of single-loop control (although the loop is closed by manual intervention). Our aim is to introduce the subject, but, more importantly, to discuss its applicability in the chemical process and, specifically, polymer industries.

In many cases of batch and semibatch polymerizations, the only measurements of the end-use properties are at the end of the batch. As discussed previously, these are often off-line, laboratory analyses. Often the same is true for the product quality variables. An example from the synthetic rubber industry is the Mooney viscosity of the rubber. Neither the Mooney viscosity (end-use property) nor its underlying product quality variables (MWD and degree of branching or cross-linking) is measured on-line. In fact, in some cases, the polymer quality variables are not measured at all on a regular basis because release specifications are written in terms of Mooney viscosity. In this situation, closed-loop control of Mooney viscosity is impossible. However, because the measurement of variables to be controlled (end use or polymer quality) are substantially uncorrelated in time, the use of SPC is appropriate. Likewise, in continuous polymerizations, the end use and polymer properties variables may be measured very infrequently and, most often, off-line. In addition, the laboratory analysis time adds a large delay to the measurement. Under these circumstances, the measurements are substantially uncorrelated in time, and, again, statistical process control may be an appropriate approach to control of polymer quality or end-use properties. Good reviews of the applicability of SPC to chemical processes are given by MacGregor [9,10].

The philosophy of SPC is to monitor the output of a process and determine when control action is necessary to correct deviations of the output from its setpoint. The most common tool for accomplishing this is the Shewhart (X-bar) chart shown in Fig. 8.7. In the discrete parts manufacturing industries, multiple samples are taken at fixed intervals. Quality tests are run on these samples, and the mean is plotted on one Shewhart chart, and the range on another. In the absence of a disturbance, the means should be normally distributed around the setpoint. If the upper and lower control limits (UCL and LCL, respectively) are placed at three standard deviations above and below the target, a range is defined into which all of the means should fall. The likelihood of a point outside the control limits means the process is "out of control," and some adjustment should be made to the process. If the process is "in control" (points are within the control limits), no action is taken. This is to prevent manipulations of the process based on stochastic variations in the process. Responding to stochastic variations in the process corresponds, in the field of conventional process control, to trying to compensate for noise; something no good control engineer would do. By definition, noise is either stochastic or deterministic, but of such high frequency that it is impossible to compensate for it. An example of this is the noise on a tank level signal coming from a differential pressure transducer which senses the hydrostatic head at the bottom of the tank and, hence, the depth of liquid. The level signal will contain a stochastic portion resulting from sensor imperfections and corruption of the signal during transmission. The level signal will also contain some deterministic information about the variations in level brought on by the turbulent nature of the liquid surface under agitation. This deterministic component is "real," but is of a very high frequency and cannot be compensated. Control engineers deal with noisy signaling by filtering inputs. This is effective when the signal is analog, or digital with high sampling rates. However, at very low sampling rates where SPC is applicable, filtering is not appropriate (due to the assumption of lack of correlation between samples) or effective. This is the rational for taking action only when the process exceeds the upper or lower control limits.

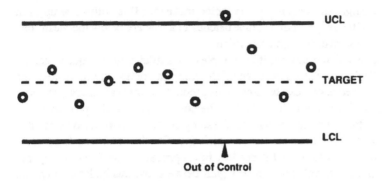

Figure 8.7   Shewhart chart for statistical process control.

Another reason for taking control action only when the process is out of control is because there is assumed to be a cost associated with control action. This is often true in the discrete manufacturing industries where it may be necessary to shut down the assembly line to make adjustments to equipment. Often this is not true in the chemical process industries where corrective action may be "free," as in the adjustment of the temperature of a polymerization reactor.

If the hypothesis is made that the points on the Shewhart chart are normally distributed, the probability of a point falling outside the control limits can be calculated, and is quite small. The likelihood of various other occurrences [9 points in a row all below (or above the mean), 14 points in a row alternating up and down, etc.] can be calculated as well. If the likelihood of these patterns is slight, they can be used as additional rules to determine when the process is out of control. Such pattern-based rules are easy for a chemical plant operator to implement.

In a batch polymerization, each batch may take up to a full day to process. In this case, it is not possible to take multiple samples and calculate the average for an X-bar chart. (It is possible to take multiple samples from each batch, but the variance will reflect only the variance of the sampling and analysis techniques, and not batch-to-batch variations.) In this situation, an "individuals chart" is used instead [11]. The individuals chart is plotted in a manner similar to the X-bar chart. The mean and standard deviation of the process are calculated and used to plot the target (setpoint or mean), UCL and LCL. In addition, a range chart is plotted. The mean of the range chart is plotted in advance. The UCL is calculated in advance. There is no LCL for the range chart. Each measurement is plotted on the individual sample chart, and the range to be plotted for each measurement is the current measurement minus the previous measurement. The process is then judged to be in control or out of control based on various patterns as above. Thus, if an individual sample for a single batch of polymer is above the UCL or below the LCL, one might suspect an error in recipe makeup. (This with show up even more readily on the range chart.) If nine batches in a row are below the mean, one might suspect raw materials contamination.

In continuous polymerization, if the samples are sufficiently infrequent that the process "settles" between samples, and if the samples are not autocorrelated, the procedures outlined above can be used. This amounts to manual steady-state control with the need for control identified by the control chart. If the process is fast compared with the sampling interval, but the samples are autocorrelated (as is often the case), a controller can be developed which specifies the correction to be made to the process. A process model is needed. For a process which is fast compared with the sampling interval, this can be a gain between the manipulated input and the process output. In addition, a disturbance model is necessary. A simple but effective model for continuous process systems is that of an integrated white noise sequence. With these assumptions, a minimum variance controller may be derived [9]. This takes the form of the discrete pure integral controller:

$$u_t = -\frac{1-\theta}{g} \sum_{j=-\infty}^{t} Y_j \qquad (8.34)$$

Here u_t is the manipulation to be made at time j, g is the process gain, Y_j is the measurement at time j, and θ is a tuning parameter. Many continuous polymerization processes fit this situation in which the process is fast with respect to the sampling interval, but the samples are still autocorrelated, and an integrated white noise model is appropriate. Other noise models may be used. If the process is not fast with respect to the sampling time, the process model must capture the dynamics of the process. If the process is "slow" with respect to the sampling time, then conventional process control is appropriate.

8.4 Applications to Polymerization Reactors

Single-input single-output (SISO) control of polymerization reactors is not commonly discussed in the literature. This is partly due to the strongly interactive nature of the polymerization system. In practice, however, polymerization reactor temperature control is often carried out as a SISO PID loop by manipulating the cooling jacket temperature to adjust reactor temperature. Other schemes for single-loop control have been proposed [12,13]. This section will discuss two application of single-loop control: the control of reactor temperature during batch polymerization and the control of monomer conversion during continuous emulsion polymerization.

8.4.1 Temperature Control in a Batch Polymerization Reactor

It is often desirable to operate a polymerization reactor isothermally. Even if isothermal operation is not critical, it may be necessary to control reactor temperature for

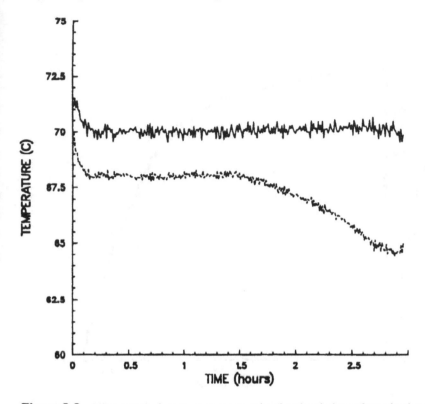

Figure 8.8 PID control of reactor temperature in a batch solution polymerization reactor. (− reactor temperature; --- jacket temperature.) (From Ref. 12.)

safety reasons (to prevent thermal runaway). Reactor temperature is often controlled by manipulating the temperature or flow rate of the coolant in the reactor cooling jacket. This scheme is hindered by the slow dynamics of heat removal and the nonlinear nature of the heat evolution process (especially in the gel effect region of free radical polymerization). If this strategy is used, it is often necessary to detune the controller in order to not become overly aggressive during times of high process gain. Some control performance can be recovered by employing a cascade strategy in which the reactor temperature (master) controller sets the setpoint for coolant temperature. The coolant temperature is then controlled at this setpoint by a slave controller which manipulates coolant flow rate into the cooling jacket.

An example of PID temperature control of a batch free radical polymerization reactor is given by Houston [14, 15]. The system is that described in Subsection 3.4.3. PID control of reactor temperature by manipulating the cooling jacket temperature is simulated in Fig. 8.8. Note that, with the exception of the initial reactor

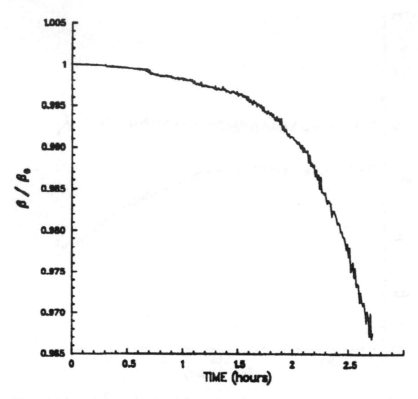

Figure 8.9 Change in process gain during polymerization. The ratio β/β_0 is the ratio of the process gain (effect of coolant jacket temperature on reactor temperature) at any time to its initial value. (From Ref. 12.)

start-up (during which all of the intitator is added), temperature control is satisfactory. (Noise has been added to the temperature measurements, which accounts for the noise level in the controlled temperature.) Figure 8.9 demonstrates the nonlinearity of the process. The ratio of the process gain (effect of jacket temperature on reactor temperature) at any time (β) to its initial value (β_0) is plotted versus time. For this particular system, the change in process gain is not excessive; this explains the good quality of control attainable with a single PID controller. For other operating conditions, variations in the process gain may necessitate the use of an adjustable gain controller. In this case, the controller gain can be specified (off-line) as a function of monomer conversion, or an adaptive controller can be used as shown in Chapter 9. Figure 3.1 shows the remaining reactor states during the control simulation in Fig. 8.8. Note the variation in MW over the course of the batch.

8.4.2 Conversion Control in a Continuous Emulsion Polymerization Reactor

In the commercial manufacture of polymers by continuous emulsion polymerization, perhaps the primary concerns are maintenance of uniform product quality and avoidance of production of poor quality material resulting from swings away from the steady-state levels of the process variables. Therefore, a control system which provides tight regulatory control of the process is a need of the industry. Design of such a system for many continuous emulsion polymerized monomers is complicated by the occurrence of the steady-state limit cycle in the number of polymer particles produced and in the monomer conversion achieved in the reaction system. In this application, control strategies are developed for the regulatory control of a series of continuous emulsion polymerization reactors and implemented in simulation of the vinyl acetate system [16].

Background

The two prominent variables to be controlled in a continuous polymerization system are the reaction temperature and the monomer conversion achieved in the reaction system. Final polymer properties are directly influenced by changes in these process variables. Several control systems for continuous emulsion polymerization have been suggested [17–20].

Most common continuous emulsion polymerization systems require isothermal reaction conditions and provide for conversion control through manipulation of initiator feed rates. Typically, as shown in Fig. 8.10, flow rates of monomer, water, and emulsifier solutions into the first reactor of the series are controlled at levels prescribed by the particular recipe being made, and the reaction temperature is controlled by changing the temperature of the coolant in the reactor jacket. Manipulation of the initiator feed rate to the reactor is then used to control reaction rate and, subsequently, exit conversion. An apparent deadtime exists between the point of addition of initiator and the point where conversion is measured. In many commercial systems, this deadtime is of the order of several hours, presenting a problem which standard PID loops are incapable of solving.

Several control techniques have been developed to compensate for a large deadtime in processes and have been reviewed by Gopalratnam et al. [21]. Among the most effective of these techniques, and the one which appears to be most readily applicable to continuous emulsion polymerization is the analytical predictor method of deadtime compensation (DTC) (Doss and Moore [22]). The analytical predictor is implemented in this application to control monomer conversion in a train of continuous emulsion polymerization reactors.

The analytical predictor, as well as the other deadtime compensation techniques, requires a mathematical model of the process for implementation. The block diagram of the analytical predictor control strategy, applied to the problem of

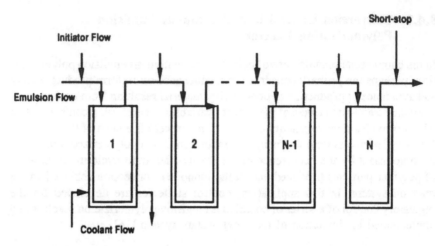

Figure 8.10 Typical continuous reactor train for emulsion polymerization. (From Ref. 13.)

conversion control in an emulsion polymerization, is illustrated in Fig. 8.11a. In this application, the current measured values of monomer conversion and initiator feed rate are input into the mathematical model which then calculates the value of conversion T' units of time in the future assuming no changes in initiator flow or reactor conditions occur during this time. Here, T' is the sum of the process deadtime, θ_d, and one-half of the sampling time, T_s,

$$T' = \theta_d + \frac{1}{2} T_s \tag{8.35}$$

(The dynamic effect of sampling is equivalent to that of a pure deadtime of one-half the sampling time). The model-predicted future value of conversion is then compared with the conversion setpoint to generate an error signal which is used by the control algorithm to achieve the new initiator feed rate setpoint. Utilizing this strategy, the control scheme for the first reactor of a series appears in Fig. 8.11a, where temperature is maintained by manipulating the reactor jacket temperature and exit conversion is controlled by means of the analytical predictor algorithm. For comparison, the block diagram depicting the standard feedback control strategy in which the initiator feed rate setpoint is manipulated by the primary conversion controller is shown in Fig. 8.11b.

Kiparissides et al. [23] detail a mathematical model for the continuous polymerization of vinyl acetate in a single CSTR. Operating conditions were shown to exist in which either steady-state operation or sustained conversion oscillations would occur for vinyl acetate. Experimental results for both cases were

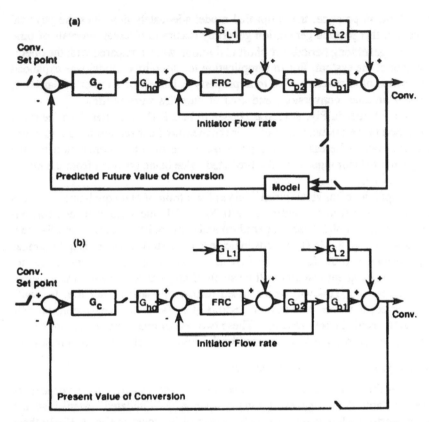

Figure 8.11 Control schemes for conversion control. (a) Block diagram of conversion control loop utilizing the analytical predictor technique of deadtime compensation; (b) conventional feedback control loop (G_{p1}) reactor process transfer function; G_{p2}: initiator flow transfer function; G_{L1}: initiator flow load; G_{L2}: reactor load; G_{h0}: zero-order hold. (From Ref. 16.)

successfully simulated by their model. Using this model, simulations have been carried out for vinyl acetate emulsion polymerization in a CSTR train, under control of the analytical predictor utilizing initiator flow rate as the manipulated variable.

Simulation Model

Kiparissides et al. [23] developed mathematical models of two levels of sophistication for the vinyl acetate system: a comprehensive model that solves for the age distribution function of polymer particles and a simplified model which solves a series of differential equations assuming discrete periods of particle nucleation. In practice, the simplified model adequately describes the physical process in that particle generation generally occurs in discrete intervals of time and these generating periods are short in duration when compared with the operation time of the system. For this application the simplified model was expanded for a series of m reactors. The total property balances for number of particles, polymer volume, conversion, and area of particles were written.

The controller in Fig. 8.11 was chosen to be a PID controller. The feedback value used in the calculation of error is the measured conversion for the standard feedback loop, whereas it is the predicted value of future conversion for the analytical predictor algorithm, the predicted value being obtained from the model of Kiparissides.

The open-loop conversion-time curves for a train of two equal-sized reactors with an initiator feed concentration of 0.005 mol/L and surfactant feed concentrations of 0.06 mol/L (stable operation) and 0.01 mol/L (sustained oscillations) are shown in Fig. 8.12. The results indicate that the second reactor is quite similar in performance to the first. If the first reactor is brought under control, the second and subsequent reactors will be stable if no particle nucleation takes place in these reactors. This will be discussed further in a later subsection. It should be observed, however, that in both regions of surfactant concentration, particle formation occurs in both reactors. These two sets of operating conditions (open-loop stable and oscillatory) are used in the following closed-loop simulations.

Conversion Control of the First Reactor

Values for the PID tuning parameters were selected by trial and error and are given in Table 8.1. For these tuning constants, the simulated system response during start-up while under closed-loop control is shown in Fig. 8.13. In these simulations, the initiator flow rate was held constant for three residence times of the reactor, at which point control action was initiated. Figure 8.13 illustrates the conversion profile in the first reactor at the low emulsifier level (0.01 mol/L). Despite prediction of the future occurrence of a particle generation, these control algorithms were not capable of preventing or even severely dampening particle nucleation at the low emulsifier concentration. It is apparent from these simulations that reactor control in the regions of low emulsifier concentration is very poor, even with deadtime compensation. These results clearly demonstrate the difficulty of reactor control in the presence of sporadic particle nucleations. Prevention of formation of new particles appears to be impossible by means of manipulation of initiator feed rate only. The free soap area in the first reactor oscillates with the conversion. The successful control strategy would provide a constant free soap area and, hence, constant particle generation in the reactor. Because this reactor is being run under a "soap-starved" condition, the free soap area is normally negative. It might be possible to start up the reactor under conditions that resulted in a constant particle generation rate and then slowly decrease the

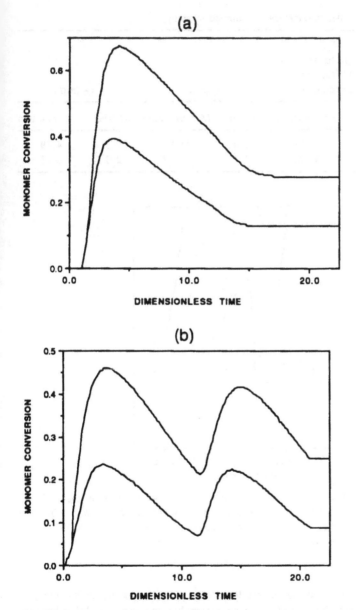

Figure 8.12 Simulated open-loop conversion of the vinyl acetate system. Initiator concentration of 0.005 mol/L H_2O and 50°C: (a) $S = 0.06$ mol/L; (b) $S = 0.01$ mol/L. (From Ref. 16.)

Table 8.1 Controller parameters for simulation study

Control scheme	K_{c1} (lb/hr or °C/fraction conversion) by trial	IAE optimum	τ_I (min) by trial	IAE optimum	τ_D (min) by trial	IAE optimum
Standard	8.0	1.68	1.0	16.83	2.0	4.29
Feedback	8.0	37.21	1.0	3.79	2.0	0.11

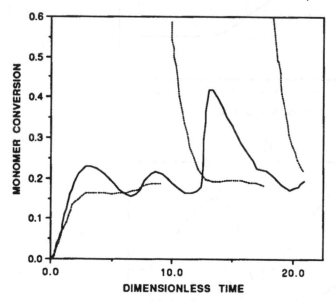

Figure 8.13 Simulated start-up of vinyl acetate polymerization at a low emulsifier level. Low surfactant (0.01 mol/L H_2O) operation under closed-loop control with arbitrarily selected controller tuning constants and manipulation of initiator flow rate at 50 °C: conversion in R1; STD feedback (−) versus DTC (---). (From Ref. 16.)

feed soap concentration to the desired level without initiating the limit cycle performance. However, this would not be an acceptable operating condition because the introduction of any disturbance into the process would be likely to send the system into the limit cycle condition. For this operating region, therefore, elimination of the unsteady-state condition by means of control strategy appears unlikely and a solution of the problem requires a design modification that would result

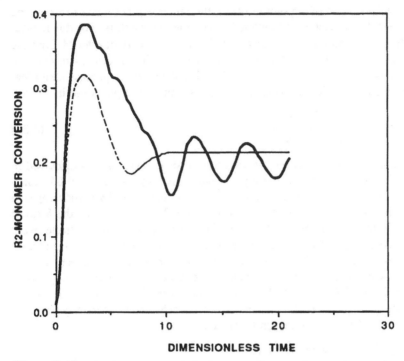

Figure 8.14 Simulated start-up of vinyl acetate polymerization at a high emulsifier level. High emulsifier level (0.06 mol/L H_2O) operation under closed-loop control with arbitrarily selected controller tuning constants and manipulation of initiator flow rate at 50 °C: conversion in R1; STD feedback (−) versus DTC (—). (From Ref. 16.)

in a constant particle generation rate in the system. This will be discussed in Chapter 9.

At high emulsifier concentration, the utility of the analytical predictor is much more apparent. Under these operating conditions, the first two reactors reach a steady-state level of polymer particle generation rate and monomer conversion. Using the same tuning constants, the simulated system response during start-up is shown in Fig. 8.14. The conversion profile for both the standard feedback and DTC algorithms manipulating initiator flow rate are shown. At the higher level of emulsifier, the free soap area is available at steady state, resulting in a constant particle generation rate. The inherent advantages of deadtime compensation under this operating condition are faster approach to setpoint and tighter regulatory control. The DTC algorithm requires smaller and less frequent changes in the manipulated variable.

Typically, instead of setpoint changes, the regulatory problem of responding to a system disturbance is encountered in commercial reactors. For this reason, the optimum tuning constants for the PID controller were developed from the IAE relations for load disturbances. First, however, it is necessary to obtain a process model of the system. Brantley [24] has developed a process identification technique which fits process data to the second-order plus deadtime form:

$$G_p(s) = \frac{K_p e^{-\theta_d s}}{(\tau_1 s + 1)(\tau_2 s + 1)} \tag{8.36}$$

The identification of a transfer function model for this system was discussed in Chapter 7. For the continuous polymerization of vinyl acetate at 0.06 mol/L emulsifier concentration, the model parameters as determined by introducing step changes into the mechanistic model are shown in Table 8.1. Given the form of Eq. (8.36), the "optimum" controller tuning constants for a PID controller were determined by the method of Gallier and Otto [25], which minimizes the integral of the absolute error (IAE) that results following a process disturbance. The effective time delay, T', used in calculating the optimum tuning constants as shown by Meyer et al. [26] is $\theta_d + T_s$ for the standard feedback loop and T_s for the analytical predictor loop. The degree of control achieved by these algorithms is measured by the common integral criteria:

1. Integral of the square error (ISE):

$$\text{ISE} = \int_0^\infty (e(t))^2 \, dt \tag{8.37}$$

2. Integral of the absolute error (IAE):

$$\text{IAE} = \int_0^\infty |e(t)| \, dt \tag{8.38}$$

3. Integral of time multiplied by the absolute value of the error (ITAE):

$$\text{ITAE} = \int_0^\infty t(e(t)) \, dt \tag{8.39}$$

where $e(t)$ is the difference between the setpoint value and the actual value of conversion.

These "controllability" criteria are used to judge the performance of the various control systems described in Table 8.2. As may be seen, a deadtime compensation markedly improves system performance at the high emulsifier concentration. A system load disturbance was simulated by introducing a step change in

Table 8.2 Comparison of control system performance—first reactor.

Manipulated variable	Surfactant conc. (mol/L)	Control method[a]	K_c	τ_I	τ_D	ISE	IAE	ITAE
			Setpoint change					
None	0.01	None[b]	—	—	—	2.123	26.03	7608
None	0.01	None[c]	—	—	—	3.518	35.46	9552
Initiator flow	0.01	SFB	8.0	1.0	2.0	2.868	24.73	7703
Initiator flow	0.01	DTC	8.0	1.0	2.0	19.65	50.51	15520
Initiator flow	0.06	None[b]	—	—	—	3.774	38.99	9512
Initiator flow	0.06	None[c]	—	—	—	5.987	36.72	4595
Initiator flow	0.06	SFB	8.0	1.0	2.0	2.123	22.78	3309
Initiator flow	0.06	SFB	1.08	16.83	4.29	2.010	19.30	1912
Initiator flow	0.06	DTC	8.0	1.0	2.0	0.397	9.11	1725
Initiator flow	0.06	DTC	37.21	3.79	0.11	0.999	10.65	1080
			System disturbance					
Initiator flow	0.06	SFB	1.68	16.83	4.29	0.0952	3.548	305.4
Initiator flow	0.06	DTC	37.21	3.79	0.11	0.0399	1.693	83.78

[a]None = open loop process; SFB = standard feedback control; DTC = deadtime compensation.
[b]Initiator concentration = 0.005 mol/L.
[c]Initiator concentration = 0.01 mol/L.

the propagation rate constant, k_p, of -5% after the system was at steady state under closed-loop control. This type of disturbance in which the system reaction rate changes suddenly, in most cases due to the introduction of an unknown disturbance, is common in commercial processes.

These simulated results for the high emulsifier concentration operating condition demonstrate the utility of deadtime compensation to the control of conversion from the first reactor in a train. With implementation of this degree of control on the first reactor, control schemes for downstream reactors can be simplified as discussed in the next subsection.

Conversion Control of Downstream Reactors

With the first reactor of the train under closed-loop conversion control, response of downstream reactors to changes in initiator flow can be approximated very closely by the second-order plus deadtime model of Eq. (8.36) and adequately controlled provided that particle nucleation in the downstream reactors occurs at a constant rate or is totally precluded because of low free soap concentration in the feed to the reactor. A danger of operating these reactor trains at high emulsifier concentration to provide stability is that, although the front reactors do reach a steady-state level of particle nucleation rate, there may exist the condition in a downstream reactor that leads to sporadic nucleations and, hence, oscillations in conversion. The required condition for stability, then, would be that the feed to the last reactor would allow for continuous particle nucleation, i.e., that positive free soap area be present in the finished emulsion. For many systems, this requirement forces a level of emulsifier in the feed that results in prohibitively high emulsion viscosities. The solution to this problem again requires a reactor design modification which provides separation of the regimes of particle nucleation and particle growth.

Conclusions

The utility of the analytical predictor method of deadtime compensation to control of conversion in a train of continuous emulsion polymerizers has been demonstrated by simulation of the vinyl acetate system. The simulated results clearly show the extreme difficulty of controlling the conversion in systems which are operated at "soap-starved" conditions. The analytical predictor was shown, however, to provide significantly improved control of conversion as compared to standard feedback systems in operating regions that promote continuous particle formation. These simulations suggest the analytical predictor technique to be the preferred method of control when it is desired that only one variable (preferably initiator feed rate) be manipulated.

References

1. P. B. Deshpande and R. H. Ash, *Computer Process Control with Advanced Control Applications*, 2nd ed., Instrument Society of America, Research Triangle Park, NC, 1988.

2. E. B. Dahlin, "Designing and Tuning Digital Controllers," *Inst. Contr. Syst.*, *41* (6 June), 1968.
3. C. E. Garcia and M. Morari, "Internal Model Control 1. A Unifying Review and Some New Results," *Ind. Eng. Chem. Proc. Des. Dev.*, *21*, 308–323 (1982).
4. G. R. Arulalan and P. B. Deshpande, "Simplified Model Predictive Control," *Ind. Eng. Chem. PRoc. Des. Dev.*, *25*, 2 (1987).
5. V. K. Chawla and P. B. Deshpande, "A New Algorithm for Single-input Single-Output Systems," *Hydrocarbon Processing*, (June 1988).
6. O. J. M. Smith, "On Closer Control of Loops with Dead-time," *Chem. Eng. Progr.*, *53*, 5, 217–219 (1957).
7. C. F. Moore, "Selected Problems in Design and Implementation of Direct Digital Control," Ph.D. thesis, Louisiana State University, Baton Rouge, 1969.
8. P. B. Deshpande, "Improve Quality Control On-Line with PID Controllers," *Chem. Eng. Progr.*, *88*, 5, 71–76 (1992).
9. J. F. MacGregor, "On-Line Statistical Process Control," *Chem. Eng. Progr.*, 84,10, 21–31, 1988.
10. J. F. MacGregor, "A Different View of the Funnel Experiment," *J. Quality Contr.*, 22, 255–259 (1990).
11. W. H. McNeese and R. A. Klein, *Statistical Methods for the Process Industries*, Marcel Dekker, Inc., New York, 1991.
12. B. Hopkins and G. H. Alford, "Temperature Control of Polymerization Reactors," *Instrum. Technol.*, 39–43 (May 1973).
13. J. F. MacGregor, A. Penlidis, and A. E. Hamielec, "Control of Polymerization Reactors: A Review," *Polym. Process Eng.*, 2, 179–206 (1984).
14. W. E. Houston and F. J. Schork, "Adaptive Predictive Control of a Semibatch Polymerization Reactor," *Polym. Process Eng.*, 5, No. 1, 119–147 (1987).
15. W. E. Houston, M.S. thesis, Georgia Institute of Technology, Atlanta, GA, 1986.
16. K. W. Leffew and P. B. Deshpande, in *Emulsion Polymers and Emulsion Polymerization* (D. R. Basset and A. E. Hamielec, eds.), ACS Symposium Series, No. 165, American Chemical Society, Washington, D.C., 1981), p. 533.
17. D. A. Wismer and W. Brand, Proceedings of the Joint Automatic Control Conference, Stanford, CA, 1964, pp. 147–154.
18. C. Kiparissides, J. F. MacGregor, and A. E. Hamielec, *AIChE J.*, *27*, 13 (1981).
19. M. J. Pollock, J. F. MacGregor, and A. E. Hamielec, in *Computer Applications in Applied Polymer Science* (T. Provder, ed.), ACS Symposium Series, No. 197, American Chemical Society, Washington, D.C., 1981, p. 209.
20. J. F. MacGregor, A. Penlidis, and A. E. Hamielec, *Polym. Process Eng.*, 2, 179 (1984).
21. P. C. Gopalratnam, P. B. Deshpande, and R. H. Ash, paper presented at ISA National Conference, Chicago, IL, Paper No. C. I. 79–619 (1979).
22. J. E. Doss and C. F. Moore, "The Discrete Analytical Predictor – A Generalized Dead Time Compensation Technique," *ISA Trans. 20*, 77–85 (1982).
23. C. Kiparissides, J. F. MacGregor, and A. E. Hamielec, *J. Appl. Polym. Sci.*, *23*, 401–418 (1979).
24. R. O. Brantley, M. S. Thesis, University Louisville, Louisville, KY, 1981.
25. P. W. Gallier and R. E. Otto, *Instrum. Technol.*, *15* (2), 65–70 (1968).
26. C. Meyer, D. E. Seborg, and R. K. Wood, *Ind. Eng. Chem. Process Des. Dev.*, *17*, (1) (1978).

9

Multivariable Control of Polymerization Reactors

To achieve optimal control of a polymerization reactor means that all the reactor outputs are maintained within specifications. Automatic control of these outputs requires a suitable multivariable strategy. In this chapter, we consider control systems design procedures for multivariable systems and their application to polymerization reactors. The treatment is based on linear principles.

Control systems design of multivariable systems begins with interaction analysis and variable pairing. The objective here is to determine the extent of interaction present in the multivariable system and how the controlled and manipulated variables should be paired. If interaction is found to be modest, multiloop control structures may be employed; the algorithms considered in Chapter 8 may then be used as control elements in each of the loops. If interaction is found to be significant, then there are two choices. In one approach, explicit decouplers may be designed for interaction compensation, whereas in the other, a full multivariable controller having interaction compensation and constraint handling capabilities may be employed. There are a number of such multivariable control techniques available [1] but we limit our discussion in this chapter to model-based control methods of the DMC (dynamics matrix control) type. Because the design procedures discussed in this chapter are based on the assumption of linearity, the resulting controllers should be tested on a full nonlinear simulation of the reactor control system prior to implementation.

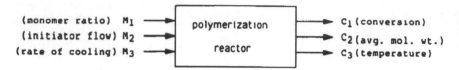

Figure 9.1 Block diagram of a typical polymerization reactor.

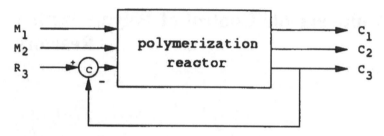

Figure 9.2 Temperature loop stabilized by conventional (proportional) controller.

Many of the multivariable control systems design procedures require process models for implementation. Such models may be developed in accordance with the procedures discussed in Chapter 7, but because many polymerization reactions are exothermic, special precautions must be taken during testing. Consider the block diagram of a typical polymerization reactor system shown in Fig. 9.1. The inputs and outputs shown are generic, selected for illustrative purposes only; variation in real-world applications will be encountered depending on the type of polymerization involved and the properties that must be controlled. As we have seen in Chapter 7, the experimental identification procedure involves perturbing the inputs and recording the outputs. However, in an exothermic reactor, it may not be possible to move the manipulated variable which denotes the rate of cooling in a stepwise fashion because it may be related to the reactor temperature by an open-loop unstable transfer function. To circumvent this problem, a proportional controller may be inserted in the feedback path giving rise to a new multivariable system shown in Fig. 9.2. The new system may be considered to be open-loop stable. The manipulated variables of the new system are M_1, M_2, and R_3 instead of M_1, M_2, and M_3 of the original system. The controlled variables remain the same.

Open-loop instability in reaction systems arises owing to the presence of multiple steady states. In such cases, a single value of rate of cooling yields three values of temperatures and conversion as shown in Fig. 9.3. These values correspond to an upper stable steady state, a lower stable steady state, and a middle unstable steady state. If an attempt is made to operate a reactor at the middle unstable

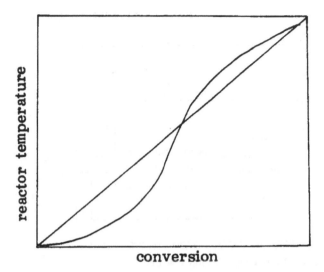

Figure 9.3   Multiple steady states in a polymerization reactor.

steady state, in the absence of feedback, the slightest bit of noise in the system will drive the system to one of the stable steady states. The trajectories that the system will traverse are entirely unpredictable. While this points to the need for feedback, it is difficult to prove unequivocally whether or not an existing commercial reactor is indeed being operated at the middle unstable steady state. Needless to say, one must exercise proper caution. For a more detailed discussion of multiple steady states in reaction systems, the reader is referred to Uppal et al. [2].

With the potential dangers due to open-loop instability having been taken care of, the traditional identification techniques such as step testing or PRBS testing may now be applied to the process depicted in Fig. 9.2, and linearized process models can be developed. Interaction analysis can then be carried out and, depending on the extent of interaction present, a multiloop control scheme shown in Fig. 9.4 or the multivariable control schemes shown in Fig. 9.5 may be employed. Some of the model-based algorithms require open-loop stability and the arrangement shown in Fig. 9.2 would have to be used if it is desired to use one of these algorithms for control.

In the material that follows, we begin with the methods of interaction analysis and variable pairing. Next, we take up multiloop control for systems where interaction is weak or where decoupling may actually deteriorate performance. This will be followed by explicit decoupling of multivariable systems. Finally, we describe model-based control techniques that inherently compensate for interaction

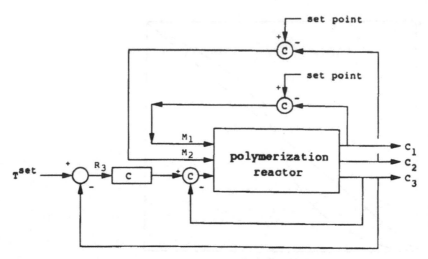

Figure 9.4 Multiloop control for polymerization reactors.

and have the ability to deal with constraints. The material presented here is brief. More details may be found in other books (e.g., Ref. 1).

9.1 Interaction Analysis

The purpose of interaction analysis is to determine the extent of interaction present in a multivariable system and also to determine how the manipulated and controlled variables should be paired. A number of methods for interaction analysis are available [3, 4]. Some use only steady-state data, whereas others use the dynamic information about the system as well. Whereas the measures use steady-state data are simple to use, they some times lead to wrong answers. Direct and inverse Nyquist arrays give complete information about interaction in a multivariable system, but their use requires access to computer-aided software [1]. In the following ,we present two methods for interaction analysis, both based on process steady-state data. Dynamic extension of each are available and the interested reader may refer to the literature for details.

9.1.1 Relative Gain Arrays

The relative gain analysis (RGA) is based on the steady-state gain matrix **K** of the multivariable system. It applies to systems having equal number of inputs and outputs. The procedure is to calculate a complimentary matrix **C** according to

$$\mathbf{C} = (\mathbf{K}^{-1})$$ (9.1)

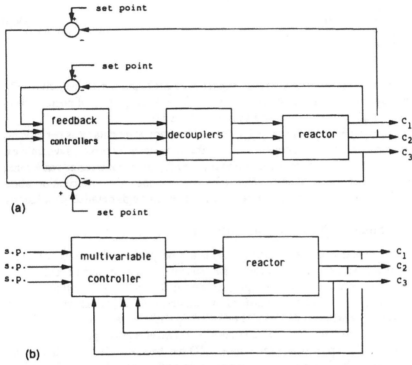

Figure 9.5 (a) Explicit decoupling for polymerization reactors; (b) Multivariable control of polymerization reactors.

The K and C matrices together give a new matrix called the relative gain array [5] λ whose elements are calculated according to

$$\lambda_{i,j} = K_{i,j} C_{i,j} \qquad (9.2)$$

A useful property of the relative gain array is that each row and each column sums to 1. Thus, in a 2×2 system, only one of the four terms need be explicitly computed. For each controlled variable, the manipulated variable selected is the one having the largest positive relative gain closest to 1. It is possible for the elements of the relative gain array to be large ($\gg 1$) or be negative. In a 2×2 system, pairings based on negative RGA elements will result in an unstable system. Also, pairings or structures having large RGA elements should be avoided. For systems of dimension higher than 2×2, Niederlinski's [6] theorem may be useful. It states that the closed-loop system resulting from the pairing $m_1 - Y_1$, $m_2 - Y_2$, ..., $m_n - Y_n$ will be unstable if

$$\frac{|K|}{\prod_{i=1}^{n} K_{ii}} < 0 \tag{9.3}$$

Thus, in a system of dimension higher than 2×2, pairing on negative RGA elements may not result in instability or, for that matter, pairing of positive RGA elements may still result in an unstable system [7].

Because the RGA is developed solely from steady-state data, it is perhaps not surprising that the conclusions it leads to are some times incorrect. Dynamic extensions of the RGA concept has been proposed [8] and the interested reader may refer to that paper for details. The RGA concept does give information about interaction and pairing that can be verified later through detailed simulations.

9.1.2 Singular Value Decomposition

Singular value decomposition [9] (SVD) is a concept from linear algebra whose properties have been exploited for multivariable control systems design. SVD analysis can be carried out with process gain matrices, as will be illustrated in this section, or with process transfer function matrices [10] which give dynamic measures of sensitivity and interaction. The geometric representation of SVD facilitates the visualization of relationships between inputs and outputs. Some researchers [3] report examples where SVD has been unable to predict proper pairings correctly. The suggested approach is to apply the variety of tools available and if any one of them indicates a potential problem, a more detailed study, including dynamic simulation, should be carried out before a final design is selected.

To use SVD based on the process gain matrix, K, it is necessary to perform the singular value decomposition of K giving three component matrices according to

$$K = U \Sigma V^T \tag{9.4}$$

where K is an $n \times m$ matrix. The matrix U is an $n \times n$ matrix whose *columns* are called the *left singular vectors*, whereas the matrix V is an $m \times m$ matrix, the *columns* of which are called the *right singular vectors*. Σ is an $n \times m$ diagonal matrix of *scalars* called *singular values*, which are organized in descending order such that $\sigma_1 > \sigma_2, > \cdots \sigma_m > 0$. The following step-by-step procedure may be used to determine U, Σ, and V.

1. Write out the transpose of the matrix K.
2. Determine the product $K K^T$.
3. Determine the eigen values, P, of $K K^T$. This requires the solution of $|KK^T - PI| = 0$.
4. σ_i is the square root (P_i)

5. Determine the matrix U by solving $(K K^T - P_1) U_1 = 0$, $(K K^T - P_2)$ $U_2 = 0$, etc., where U_1, U_2, etc., are columns of the matrix U.
6. Determine the matrix V by solving $(K^T K - P_1) V_1 = 0$, $(K^T K - P_2)$ $V_2 = 0$, etc., where V_1, V_2, etc., are the columns of the matrix V.

According to SVD rules, the pairing that will yield the least open-loop interaction is one in which the output associated with the largest magnitude element (without regard to sign) of the U_1 vector is paired with manipulated variable associated with the largest magnitude element (again, without regard to sign) of the V_1 vector. The output associated with the largest magnitude element of U_2 is paired with the manipulated variable associated with the largest magnitude element of V_2, and so on.

SVD analysis gives an important index of controllability, namely, the condition number of the system. The condition number is the ratio of the largest singular value, σ_{max}, to the smallest singular value, σ_{min}. Condition numbers are used to determine the *condition* of a set of equations; the larger the condition number, the more difficult it is to compute the matrix inverse that is free of computational errors. In terms of process control, for systems having large condition numbers, it may be impossible to satisfy the entire set of control objectives.

An attractive feature of SVD analysis is that it gives not only the condition number of the full multivariable system but also the condition number of all multivariable subsets that comprise the full system. As an example Moore [9] has computed the following condition numbers of a 4 × 4 dryer system:

Dimension	Condition Number
4 × 4	60.43
3 × 3	20.55
2 × 2	7.02

These condition numbers show on a relative basis how much more difficult it will be to control four controlled variables as opposed to only two. It must be emphasized that these conclusions have been arrived at on the basis of steady-state information only and they would need to be verified, for example, by dynamic simulation.

It has been pointed out that in most cases the SVD pairing procedure gives results that are consistent with the Relative Gain Analysis [9]. However, there are cases where SVD and the RGA lead to different conclusions.

1. RGA Problems. For systems that are characterized by high condition numbers, the SVD and RGA analyses may suggest different pairings. However,

for high condition numbers, multivariable controls are not likely to succeed regardless of controller pairing (or decoupling or another multivariable strategy). The RGA analysis by itself does not flag such conditions. The SVD analysis clearly indicates such problems, and the controller pairing under such conditions can be viewed as a systematic way to reduce the control objectives to a more reasonable level.

2. SVD Problems. The SVD analysis addresses only the open-loop nature of the system and recommends controller pairing that will give the control system the greatest open-loop advantage in terms of loop sensitivity and loop interactions. It does not address the closed-loop "decommissioning" problem that the RGA does. The RGA flags, with negative numbers, those controller pairings which would change sign if some other control loop(s) were to be put in manual. This is indeed a serious problem in the process industries where loops are frequently taken off and put back on-line.

9.2 Multiloop Control

If the interaction analysis suggests that only a modest amount of interaction is present, or that the condition number is high, then, it may be desirable to employ SISO controllers. The conceptual arrangement of SISO controllers operating in a multivariable environment was depicted in Fig. 9.4. The feedback controllers shown may involve PID type algorithms or model-based algorithms such as IMC (Dahlin), SMPC, or CMBC which were discussed in Chapter 8; the latter can be used provided the open-loop unstable elements, if any, have been stabilized by conventional feedback. In the following, we discuss the problem of finding tuning constants of multiloop PI controllers. Then we present an application of model-based control to a simulated polymerization reactor system.

9.2.1 Biggest Log-Modulus Tuning of PI Controllers

The objective is to determine the gains and integral times of these controllers such that the multivariable system gives good set point and load responses (11). It is also desirable that the system have *integrity*, meaning that the system must remain stable in the event one or more controllers are switched from automatic to manual or vice versa. In this development, it is assumed that the system is of dimension $n \times n$ and is controlled by n PI controllers such that an output C_i is controlled by an input M_i. Also, the multivariable process is assumed to be open-loop stable. The method utilizes Nyquist stability criteria. We begin with a review of the stability properties of SISO control systems. A closed-loop control system can be shown to be stable if all the roots of the characteristic equation

$$1 + G_c G_p = 0 \qquad\qquad (9.5)$$

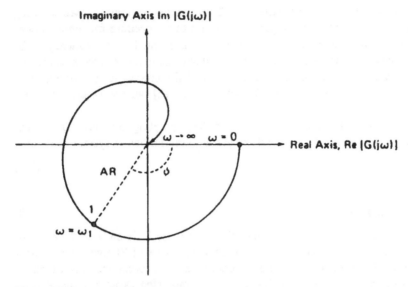

Figure 9.6 Typical Nyquist plot.

lie in the left-half s-plane. Here G_c represents the controller and G_p is the process transfer function. The Laplace transform operator has been omitted for brevity. For a given G_p, the controller gain K_c and integral time τ_I are selected such that the stability condition is met.

The Nyquist stability criteria are developed around a Nyquist plot. A Nyquist plot depicts the real part of $G(j\omega)$ on the x axis and the imaginary part of $G(j\omega)$ on the y axis. A typical Nyquist plot is shown in Fig. 9.6. A point on the Nyquist plot at a particular frequency gives the corresponding amplitude ratio and phase angle for that frequency, and, thus, Nyquist plots can be prepared from the familiar Bode plots. The Nyquist stability criterion states that a feedback control system will be unstable if the Nyquist plot of $G(s)$ encircles the point $(-1,0)$ on the negative real axis as the frequency increases from zero to infinity. The number of encirclements corresponds to the number of roots of the characteristic equation that lie in the right-half s plane assuming that the process is open-loop stable.

With the appropriate choice of K_c and τ_1, the Nyquist plot can be moved sufficiently away from the point $(-1,0)$, thus imparting a sufficient margin of stability to the control loop. A measure of the distance of the $G(j\omega)$ contour from the point $(-1,0)$ is given by

$$L_c = 20 \log \left| \frac{G}{1 + G} \right| \qquad (9.6)$$

where $G = G_c G_p$. A specification of $+2$ dB for the maximum closed-loop log modulus, $L_c(\text{max})$ has been suggested. To use this procedure for tuning a controller of an SISO system, one would plot L_c as a function of frequency ω for trial values of K_c and τ_I. If the maximum value of L_c so determined exceeds $+2$ dB, new values of K_c and τ_I are selected, and the procedure is repeated until satisfactory results are achieved. A Nichols chart [12] can also be used to determine L_c from $G(j\omega)$.

The procedure for multiple SISO controllers operating in a multivariable environment is as follows. We would begin with the following system equations:

$$C = G_p M \tag{9.7}$$

and

$$M = G_c(R - C) \tag{9.8}$$

where C is a $n \times 1$ vector of controlled variables, M is an $n \times 1$ vector of manipulated variables, G_c is an $n \times n$ diagonal matrix of PI controller transfer functions, G_p is an $n \times n$ matrix containing the process transfer functions relating the n controlled variables to the n manipulated variables, and R is an $n \times 1$ vector of setpoint values of the n controlled variables. Combining Eqs. (9.7) and (9.8) gives

$$C = (I + G_p G_c)^{-1} G_p G_c R \tag{9.9}$$

Because the determinant of a matrix winds up in the denominator in the process of evaluating a matrix inverse, the characteristic equation of a multivariable system is the scalar equation

$$\det(I + G_p G_c) = 0 \tag{9.10}$$

If the left-hand side of Eq. (9.10) is plotted against frequency, the encirclements of the origin would indicate that the system is unstable. If a new function w is defined as

$$w = -1 + \det(I + G_p G_c) \tag{9.11}$$

and plotted as a function of frequency, then the encirclement of the point $(-1,0)$ would indicate instability.

Now, synonymous with the SISO controller design procedure, a multivariable closed-loop log modulus is defined as

$$L_{cm} = 20 \log \left| \frac{w}{1 + w} \right| \tag{9.12}$$

Based on a study of 10 multivariable systems, a value of $L_{cm}(\text{max}) = 2n$ has been suggested, where n is the dimension of the multivariable system. The tuning procedure for n SISO controllers is as follows:

1. Calculate Ziegler–Nichols settings of n PI controllers. To do this calculation numerically, the value of the crossover frequency, ω_{co}, for which the phase angle is exactly $-180°$ is required. The reciprocal of the gain corresponding to this frequency is the ultimate gain, K_u. Then, the Ziegler–Nichols settings may be computed according to

$$K_c = \frac{1}{2.2} K_u \qquad (9.13a)$$

and

$$\tau_I = \frac{2.0\,\pi}{1.2\,\omega_{co}} \qquad (9.13b)$$

2. Assume a factor F; typical values are said to vary between 2 and 5.
3. Calculate new values of controller constants by the relationships

$$K_{c,j} = \frac{K_{c,\text{Z.N.}}}{F} \qquad (9.14a)$$

and

$$\tau_{I,i} = \tau_{I,\text{Z.N.}}F \qquad (9.14b)$$

The factor F may be considered as a detuning factor; as the value of F increases, the stability margin increases, but then the system becomes more sluggish and vice versa. The suggested procedure is meant to give a reasonable compromise between stability and performance.

4. Compute w according to

$$w = -1 + \det(1 + G_p G_c) \qquad (9.15)$$

For a 2×2 system, for example,

$$\det(I + G_p G_c) = 1 + G_{c1}G_{11} + G_{c2}G_{22} + (G_{c1}G_{c2})(G_{11}G_{22} - G_{12}G_{21})$$

5. Determine

$$L_{cm} = 20 \log \left| \frac{w}{1 + w} \right| \qquad (9.16)$$

6. If $L_{cm}(\text{max}) > 2n$, then assume a new value of F and return to step 2.

This procedure has been called biggest log-modulus tuning (BLT). The procedure guarantees stability with all controllers in automatic, also with individual controllers in automatic and the rest in manual. Further checks may have to be made for other manual/automatic combinations. Also, this procedure may be considered as giving preliminary settings; further trial and error may be required in the field.

9.2.2 CMBC for Multiloop Control

In this section, an application of the model-based CMBC algorithm to a simulated methyl methacrylate polymerization reactor is presented [13]. Recall that the CMBC algorithm is given by

$$M(z) = \frac{1 - \beta z^{-1}}{1 - \beta} \frac{1}{K_p} \left[E(z) + \sum_{i=1}^{N} h_i z^{-i} M(z) \right] \qquad (9.17)$$

where h_i are the impulse response coefficients and β is the single tuning constant of this algorithm.

Solution polymerization of methyl methacrylate in a CSTR has been modeled based on known free radical polymerization kinetics. The nonisothermal plant model consists of mass balances over monomer, initiator, and solvent coupled with an energy balance. The polymerization model also incorporates a "gel-effect" correlation to account for the changes in reaction kinetics that occur at high conversions, due to the increase in solution viscosity. The plant simulation uses a set of dimensionless modeling equations. Details of the model derivation and the model equations may be obtained from Kwalik [14] and Kwalik and Schork [15]. This model is chosen because it accurately represents the nonlinearities and multiple steady states that have been observed in practice.

In this example, the controlled variables are taken to be reactor temperature, T, and monomer concentration, M, and the manipulated variables are jacket coolant temperature T_c and initiator feed concentration I. It is assumed that adequate control of molecular weight can be achieved by controlling these two variables. Two independent SISO controllers are used in the multiloop strategy; the jacket coolant temperature is used to control the reactor temperature and the initiator feed concentration is used to control the monomer concentration in the reactor. In the simulation study, the controlled and manipulated variables are monitored in terms of the dimensionless variables M^*, T^*, T_c^*, I^* for convenience of computation.

The simulated reactor has a volume of 900 L and a heat transfer area of 2.8 m^2. The feed concentrations of solvent, monomer, and initiator are 3.06, 6.57, and 0.071 gmol/L, respectively. The reactants are fed to the reactor at 320K. In the first case studied, a volumetric feed flow rate of 0.25 L/sec and a sampling time of 0.0125 residence time units have been used. The steady-state monomer concentration is 0.092 gmol/L ($M^* = 0.014$) and reactor attains a temperature of 366.23K [$T^* = 0.989$].

To implement CMBC, an impulse response model of the process is required. Such a model was derived from the mechanistic model by introducing a step change in each of the manipulated variables and recording the corresponding changes in the controlled variables. Fig. 9.7 gives the observed responses. The impulse response model of the reactor was obtained from these curves. The use of impulse

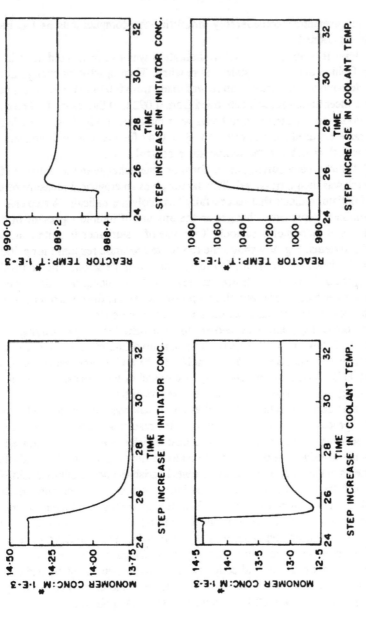

Figure 9.7 Open-loop step response of reactor (Damkohler number 123.7×10^5).

response representation considerably simplifies the computations as has been previously mentioned.

To implement control, the reactor simulation was run from cold start in an open-loop fashion until steady state was reached. Then, a setpoint change in the dimensionless monomer concentration was made from 0.014 to 0.015; the actual monomer concentration setpoint changed from 0.092 to 0.099 gmol/L. The controller smoothly tracked the setpoint change as indicated in Fig. 9.8. The CMBC tuning parameters used for this test were $\beta = 0.47$ for the monomer concentration loop and $\beta = 0.5$ for the temperature control loop.

A load disturbance occurring in the form of a step change in the initiator efficiency factor was then introduced. This factor gives an empirical measure of the fraction of initiator radicals that successfully form polymer radicals. A step change in this parameter represents changes in the amount of inhibitor present in different batches of monomer feedstock. This type of disturbance is quite common in industrial systems. Fig. 9.9 shows the reactor temperature response for a 100% change in efficiency factor, from 0.5 to 1, which is quite good. The controller brings the reactor back to the desired steady state by reducing the initiator concentration appropriately. The impulse response model and the controller settings used in this test were the same as ones used in servo control.

Another operating point was chosen for the next test corresponding to a volumetric feed flow rate of 0.4 L/sec. The start-up of the reactor led to a steady state monomer concentration of 6.15 gmol/L (M^*) and a temperature of 324K ($T^* = 0.0907$). Again, an impulse response model of the process was obtained by introducing step changes in the manipulated variables.

Now a ramp decrease in the heat transfer coefficient was used as the load disturbance to the process to simulate the fouling of the reactor walls that occurs during operation between reactor clean-ups. The heat transfer coefficient is reduced to 75% of its initial value in a period of 10 residence time units. Fig. 9.10 indicates the movements of the controlled and manipulated variables under multiloop CMBC. With less heat transfer, the reactor moves into a region of greater nonlinearity. The combination of nonlinearities and lack of decoupling results in the oscillations of initiator and monomer concentrations and the relatively poor response of the temperature loop. These responses may be improved by using multivariable controllers of the type we discuss in the ensuing sections. However, we have shown that despite the nonlinearities and coupling, this easy-to-implement algorithm has been shown to give good results for several of the tests and the choice of this type of multiloop control may well be adequate in selected applications.

9.3 Multivariable Control with Explicit Decoupling

If the relative gains are numerically close to one another, interaction ("fighting loops") in a multivariable control system is likely to be a problem, particularly

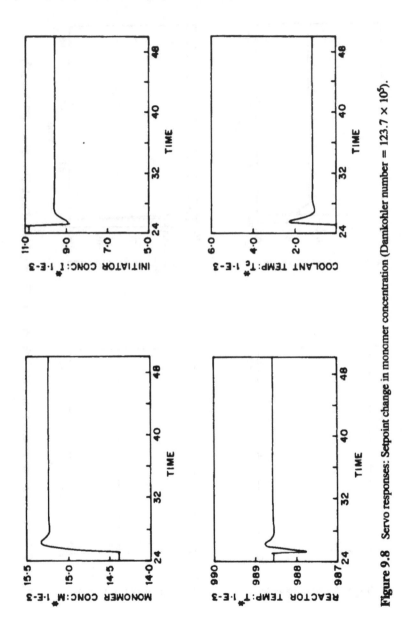

Figure 9.8 Servo responses: Setpoint change in monomer concentration (Damkohler number = 123.7×10^5).

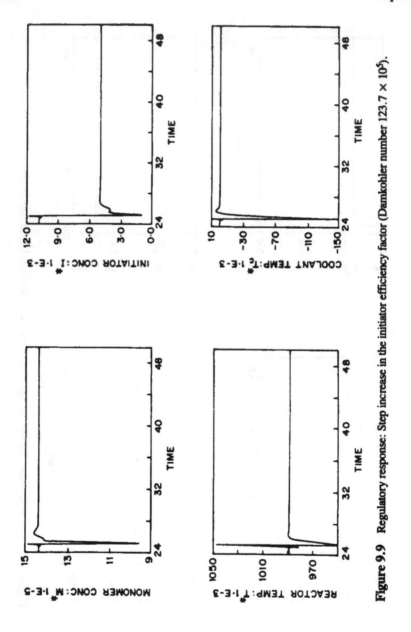

Figure 9.9 Regulatory response: Step increase in the initiator efficiency factor (Damkohler number 123.7×10^5).

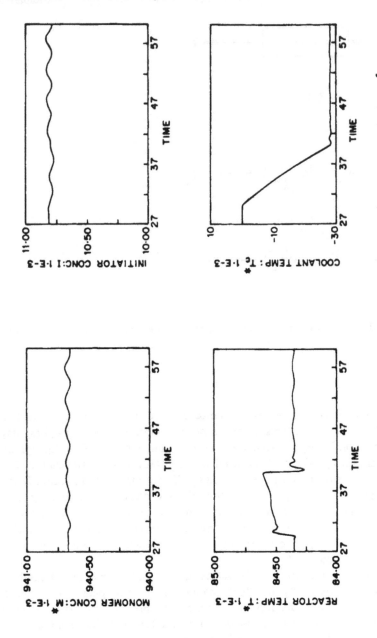

Figure 9.10 Regulatory response: Ramp decrease in heat transfer coefficient (Damkohler number = 77.3×10^5).

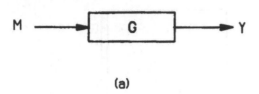

(a)

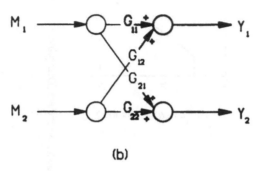

(b)

Figure 9.11 Open-loop multivariable systems. (a) A general $n \times n$ system; (b) a 2×2 example.

if the response times of the loops are comparable. In cases where cross-coupling between loops is severe, the system can become unstable, and decoupling may improve control. There are two approaches to decoupling. In explicit decoupling, a decoupler is designed such that one input affects only its associated output. The system thus decoupled may be controlled by PI or some other algorithm. In the second approach, one designs a full multivariable controller that inherently compensates for interaction and accommodates any constraints that might be present. In this section we present explicit decoupling concepts and their use in multivariable control systems. To implement a decoupler, a process model or frequency response of the process is required.

A decoupler is a device that eliminates interaction between the manipulated and controlled variables by, in effect, changing all the manipulated variables in such a manner that only the desired controlled variable changes. To begin, consider the block diagram of an open-loop multivariable system shown in Fig. 9.11. In the following derivation, z-transforms are assumed throughout, but the z-transform operator has been omitted for brevity. The open-loop pulse transfer function matrix relating the controlled and the manipulated variables for a general $n \times n$ system is

$$\begin{bmatrix} Y_1 \\ Y_2 \\ \vdots \\ Y_n \end{bmatrix} = \begin{bmatrix} G_{11} & G_{12} & \cdots & G_{1n} \\ G_{21} & G_{22} & \cdots & G_{2n} \\ \vdots & \vdots & \vdots & \vdots \\ G_{n1} & G_{n2} & \cdots & G_{nn} \end{bmatrix} \begin{bmatrix} M_1 \\ M_2 \\ \vdots \\ M_n \end{bmatrix} \quad (9.18)$$

where

$$G_{i,j} = z\{G_{h0}(s)\, G_{i,j}(s)\} \quad (9.18a)$$

Equation (9.18) may be compactly written as

$$Y = GM \quad (9.18b)$$

To achieve decoupling, a matrix of decouplers D is inserted ahead of G as shown in Fig. 9.12 such that each input to the decoupler affects only one controlled variable. Mathematically, this means that the product of D and G must be a diagonal matrix which we denote as H. Thus,

$$M = DU \quad (9.19)$$

Therefore,

$$Y = GDU = HU \quad (9.20)$$

where

$$H = \begin{bmatrix} H_1 & 0 & \cdots & 0 \\ 0 & H_2 & \cdots & 0 \\ \vdots & \vdots & & \vdots \\ 0 & \cdots & & H_n \end{bmatrix} \quad (9.20a)$$

Solving Eq. (9.20) for D gives

$$D = G^{-1}H \quad (9.21)$$

Denoting the elements of G^{-1} by the symbol $\{g_{i,j}\}$, Eq. (9.21) may be written in the expanded form as

$$\begin{bmatrix} D_{11} & D_{12} & \cdots & D_{1n} \\ D_{21} & D_{22} & \cdots & D_{2n} \\ \vdots & \vdots & \vdots & \vdots \\ D_{n1} & D_{n2} & \cdots & D_{nn} \end{bmatrix} = \begin{bmatrix} g_{11} & g_{12} & \cdots & g_{1n} \\ g_{21} & g_{22} & \cdots & g_{2n} \\ \vdots & \vdots & \vdots & \vdots \\ g_{n1} & g_{n2} & \cdots & g_{nn} \end{bmatrix} \begin{bmatrix} H_1 & 0 & \cdots & 0 \\ 0 & H_2 & \cdots & 0 \\ \vdots & \vdots & \vdots & \vdots \\ 0 & 0 & \cdots & H_n \end{bmatrix}$$

$$(9.22)$$

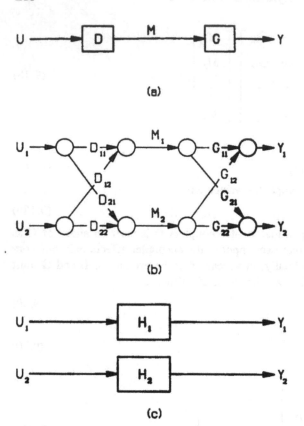

Figure 9.12 Traditional decoupling concepts.

This set has $n^2 + n$ unknowns (n^2 D's and nH's), but only n^2 equations are available to calculate them. Thus, the system is underspecified by n equations. Of these, n unknowns must somehow be fixed so that the remaining n^2 elements can be determined. This is accomplished in the traditional approach to decoupling in one of two ways. In *ideal decoupling*, it is specified that the response of each loop (with all loops on automatic control) be the same as the response one would get if the other loops were on manual (thus fixing the other manipulated variables). This allows the specification of $H_1, H_2, \ldots, H_n$ and, thus, the n^2 elements of D can be calculated. In *simplified decoupling*, the diagonal elements of D are set equal to 1, thus allowing for the off-diagonal elements of D and $H_1, H_2, \ldots, H_n$ to be computed. Although either approach can be used in decoupling, simplified decoupling is used in this section. Since $D_{11}, D_{22}, \ldots, D_{nn}$ are set equal to 1, Eq. (9.22) gives

$$D_{i,j} = \frac{g_{i,j}}{G_{j,j}} \quad \text{for } i \neq j \tag{9.23a}$$

with

$$H_i = \frac{1}{g_{i,i}} \tag{9.23b}$$

9.3.1 PID Type Control

The open-loop decoupled system may be controlled with n PID type controllers whose tuning constants may be determined on the basis of the elements of the matrix H. A potential problem with this type of control is that in the presence of a plant-model mismatch (i.e., modeling errors) it becomes difficult to tune the interacting constants of the n PID controllers. Perhaps for this reason, few successful applications of decoupled PID control have been reported in the literature.

9.3.2 Reference System Decoupling

Here, we present a newer approach to the automatic control of decoupled systems leading to a controller having one tuning constant per loop [16]. A block diagram of the control system based on reference system decoupling (RSD) is shown in Fig. 9.13. The objective is to design the controller G_c such that the multivariable system yields desired closed-loop responses. The closed-loop pulse transfer function matrix of the system in Fig. 9.13 to a change in setpoint R is

$$P = (I + GDG_c)^{-1} GDG_c \tag{9.24}$$

Solving Equation (9.24) for G_c gives

$$G_c = (GD)^{-1} P(I - P)^{-1} \tag{9.25a}$$

or

$$G_c = H^{-1}P(I - P)^{-1} \tag{9.25b}$$

Note that because the system is decoupled, P is a diagonal matrix and, thus, G_c is a diagonal matrix also, as expected. We are free to choose the matrix P as long as G_c remains realizable. If the plant transfer function matrix G contains first order with deadtime (FODT) elements, then the same form may be specified for the elements of P giving

$$P = \text{diag}\left\{ \frac{1 - \beta_1}{1 - \beta_1 z^{-1}} z^{-(N_1+1)} , \cdots , \frac{1 - \beta_n}{1 - \beta_n z^{-1}} z^{-(N_n+1)} \right\} \tag{9.26}$$

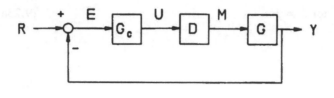

(a)

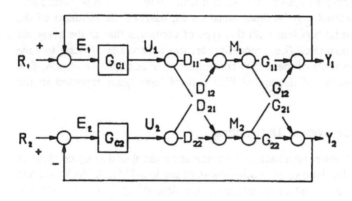

(b)

Figure 9.13 Reference system decoupling for multivariable systems.

where N_i $(i = 1, ..., n)$ represent the deadtimes (in terms of number of sampling periods) in H_i $(i = 1, ..., n)$, and β_i's determine the speed of closed-loop responses of Y_i. The β_i's lie between 0 and 1. In the absence of modeling errors, the β's can be tightly tuned. In the presence of modeling errors, they may be adjusted to maintain robustness. Larger values of β's closer to 1 make the system responses more sluggish, and vice versa.

The solution of Equation (9.25b) for G_c is

$$G_c = \text{diag}\{G_{c1}, G_{c2}, ..., G_{cn}\}$$

or

$$G_c = \text{diag}\left\{\frac{P_1}{H_1(1 - P_1)}, ..., \frac{P_n}{H_n(1 - P_n)}\right\} \tag{9.27}$$

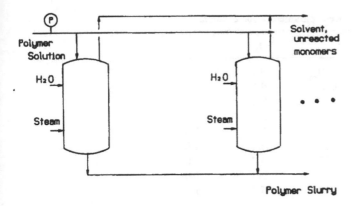

Figure 9.14 Schematic of flash drum process.

where the H_i's are given in Eq. (9.23b) and the P_i's are given in Eq. (9.26). Cross-multiplying the terms in Eq. (9.27) and inverting gives the controller outputs U_i in the time domain. The algorithm in Eq. (9.27) is based on linear control concepts. If a nonlinearity is introduced into the control system by manipulated variable saturation, it will become necessary to detune the controllers by choosing higher values of β_i or by using a separate anti- (reset) windup scheme.

As an example, we present an application involving multiple flash drums operated in parallel at an industrial plant site. A three-drum operation has been selected for the illustration of the method. The polymer solution, "cement," leaving a reactor is contacted with high-pressure steam and hot water in the flash drums, causing the solvent and the unreacted monomers to vaporize. The polymer product is recovered as crumb particles in a water slurry. The feed valves to the flash drums are used to regulate the reactor pressure and control the cement split between flash drums to produce consistent polymer crumb. A schematic of the process is shown in Fig. 9.14. Pseudo-random binary sequence (PRBS) testing in the field was employed to identify the process dynamics by time series analysis [16]. The resulting transfer functions are shown in Table 9.1(a). The corresponding pulse transfer functions are shown in Table 9.1(b). The system in the field is on multiloop PI control and it is desired to assess the suitability of RSD for this application.

For the choice of $D_{11} = D_{22} = D_{33} = 1.0$, Eq. (9.23) gives

$$D = \begin{pmatrix} 1 & -1 & -1 \\ 1 & 1 & 0 \\ 1 & 0 & 1 \end{pmatrix} \qquad (9.28)$$

Table 9.1(a)   Process parameters for example.

Process	K_p	τ_p	θ_d
G_{11}	−0.10	0.0	0.0
G_{12}	−0.10	0.0	0.0
G_{13}	−0.10	0.0	0.0
G_{21}	−0.17	5.1	1.0
G_{22}	0.34	5.1	1.0
G_{23}	−0.17	5.1	1.0
G_{31}	−0.17	5.1	1.0
G_{32}	−0.17	5.1	1.0
G_{33}	0.34	5.1	1.0

Load	K_L	τ_L	θ_L
G_{L1}	1.0	0	0
G_{L2}	1.0	0	0
G_{L3}	1.0	0	0

and

$$
\mathbf{H} = \begin{bmatrix}
-0.3z^{-1} & 0 & 0 \\[2mm]
0 & \dfrac{0.0196z^{-6}}{1+0.9615z^{-1}} & 0 \\[4mm]
0 & 0 & \dfrac{0.0196z^{-6}}{1+0.9615z^{-1}}
\end{bmatrix}
\tag{9.29}
$$

Now, the controllers can be calculated according to

$$
G_{c,i} = \frac{U_i(z)}{E_i(z)} = \frac{P_i}{H_i(1 - P_i)}
\tag{9.30}
$$

where P_i are the elements of the diagonal closed-loop pulse transfer function matrix of the type given in Eq. (9.26). Substitution for P_i and H_i in Eq. (9.30) followed by cross-multiplication and inversion gives the controller outputs in the time domain

$$
U_1(k) = U_1(k - 1) - \frac{1 - \beta_1}{0.3}E_1(k)
\tag{9.31}
$$

$$
U_2(k) = \beta_2 U_2(k - 1) + (1 - \beta_2)U_2(k - N_2 - 1)
$$

$$
+ \frac{1 - \beta_2}{0.51(1 - \alpha_2)}[E_2(k) - \alpha_2 E_2(k - 1)]
$$

Table 9.1(b) Pulse transfer functions for example.

Process	Pulse transfer function
$G_{11}(z)$	$-0.1z^{-1}$
$G_{12}(z)$	$-0.1z^{-1}$
$G_{13}(z)$	$-0.1z^{-1}$
$G_{21}(z)$	$\dfrac{-0.00654z^{-6}}{1-0.9615z^{-1}}$
$G_{22}(z)$	$\dfrac{0.01307z^{-6}}{1-0.9615z^{-1}}$
$G_{23}(z)$	$\dfrac{-0.00654z^{-6}}{1-0.9615z^{-1}}$
$G_{31}(z)$	$\dfrac{-0.00654z^{-6}}{1-0.9615z^{-1}}$
$G_{32}(z)$	$\dfrac{-0.00654z^{-6}}{1-0.9615z^{-1}}$
$G_{33}(z)$	$\dfrac{0.01307z^{-6}}{1-0.9615z^{-1}}$
$G_{L1}(z)$	z^{-1}
$G_{L2}(z)$	z^{-1}
$G_{L3}(z)$	z^{-1}

$$U_3(k) = \beta_3 U_3(k-1) + (1-\beta_3)U_3(k-N_3-1)$$

$$+ \frac{1-\beta_3}{0.51(1-\alpha_3)}\,[E_3(k) - \alpha_3 E_3(k-1)]$$

where

$$\alpha_2 = \alpha_3 = \exp\left\{\frac{-T}{\tau}\right\} = 0.9615$$

$$T = 0.2 \text{ min}$$

$$\tau_1 = \tau_2 = \tau_3 = 5.1$$

$$N_2 = N_3 = \text{int}\left\{\frac{\theta}{T}\right\} = 5$$

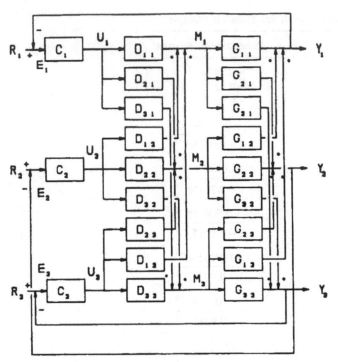

Figure 9.15 Block diagram of the RSD scheme for the 3×3 system.

and

$$\theta_1 = \theta_2 = \theta_3 = 1.0$$

A block diagram of the RSD system for this process is shown in Fig. 9.15. The following tests were conducted to assess the performance of RSD for this application.

1. The responses of RSD and decoupling PI control for two combinations of setpoint changes have been evaluated in absence of modeling errors. Figure 9.16 shows the closed-loop responses for these tests for both control strategies.

2. The responses of RSD and decoupling PI control for a unit step change in load affecting all loops have been evaluated in the absence of modeling errors. The tabular results for these tests are shown in Table 9.2 and the corresponding responses are shown in Fig. 9.17.

3. The regulatory performance of RSD and decoupling PI control in the presence of a modeling errors in each of the parameters ranging from 0 to 150% has been evaluated and the tabular results are also shown in Table 9.2. Representative plots for one of the tests are shown in Fig. 9.18.

The foregoing results show that the servo performance of RSD is similar to decoupling PI control for comparable valve movements. However, RSD is capable of providing deadbeat control in the absence of modeling errors. The regulatory performance of both strategies in the absence of modeling errors is also similar, but the tabular data in Table 9.2 shows that RSD exhibits a remarkable level of robustness vis-à-vis decoupling PI control. It may be noted that as the modeling errors were increased to 150%, PI control became unstable. RSD was stable in the presence of 150% modeling errors although the control valve movements showed a tendency toward ringing. By increasing the β's ringing could be completely eliminated at a relatively small cost in terms of ISE (integral square error).

9.4 Multivariable Control Strategies

In the context of this section, a multivariable control strategy is one that provides inherent deadtime and interaction compensation. Such strategies may be used in cases where interaction is significant enough to render multiloop control ineffective. It may be noted that if the condition number of the system is sufficiently large, then one may not have any choice but to use multiloop control. A number of multivariable control strategies are currently available [1]. They include Nyquist arrays, pole placement methods, factorization methods, state-space methods, among others. In this section, we focus our attention on model-predictive control and its application to polymerization systems.

9.4.1 Model-Predictive Control

If all the reactor outputs must be controlled in the presence of constraints, then model-predictive control may offer an effective solution. In assessing the applicability of MPC to exothermic reactors, the following points must be kept in mind.

1. Model-predictive control algorithms are derived from linear control principles, whereas the polymerization reactions can be highly nonlinear, especially if reaction conditions are changed.

2. Model-predictive control methods apply to processes that are open-loop stable. Thus, potentially open-loop unstable processes must first be stabilized as has been previously described, before MPC is applied.

Several versions of MPC algorithms are available in the market. Among the first to appear were *ADERSA*'s Model Algorithmic Control (commercially known as IdCom, *Id*entification & *Com*mand) and *SHELL*'s Dynamic Matrix Control (DMC). A version of DMC marketed by DMCC also comes with an excellent process identification package for the analysis of experimental input–output data to identify step response models that are required in the implementation of DMC. A unifying theory of model-predictive control is also available. In this section,

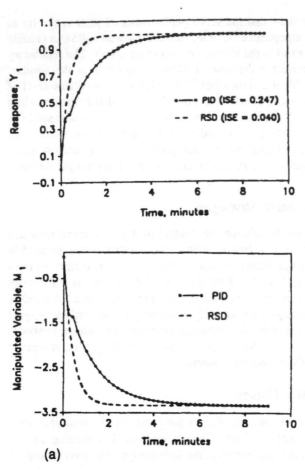

Figure 9.16 Servo responses of the 3×3 system in the absence of modeling errors: (a) Setpoint change in loop 1; (b) setpoint change in loop 2.

we present the DMC version of model-predictive control strategy for SISO (single-input single-output) and MIMO (multiple-input multiple output) systems.

For simplicity we begin with single-input single-output systems first. The block diagram of a DMC system is shown in Fig. 9.19. Model-predictive control systems utilize a mathematical model of the plant to predict future outputs, correct them for the presence of disturbances and/or modeling errors, compare the corrected outputs with the setpoints to generate a vector of projected errors, and use this vector to calculate controller outputs. The detailed procedure is described in the following paragraphs.

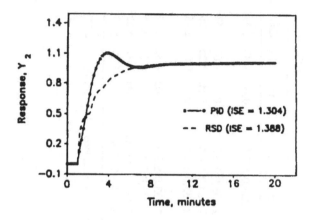

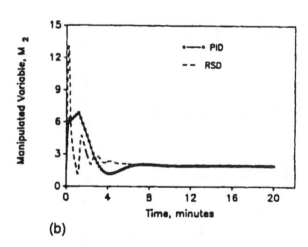

(b)

Figure 9.16 Continued

Model Prediction

In DMC, a step response model is used for prediction. Introducing the step response coefficients $(a_i - a_{i-1})$ for h_i into Eq. (7.42) followed by rearrangement gives the one-step ahead predictor according to

$$O_{k+1} = \sum_{i=1}^{k+1} a_i \Delta I_{k+1-i} + O_s \tag{9.32}$$

Table 9.2 Regulatory responses for example.

Load disturbance	Modeling error (%)	Decoupled PI			Controller $(\beta_1, \beta_2, \beta_3)$	RSD		
		ISE_1	ISE_2	ISE_3		ISE_1	ISE_2	ISE_3
1,1,1	0	0.25	1.30	1.30	2.333,0.5,0.5	0.04	1.39	1.39
1,1,1	40	0.13	2.04	2.04	2.333,0.5,0.5	0.01	1.67	1.67
1,1,1	80	0.08	4.02	4.02	2.333,0.5,0.5	0.00	2.25	2.25
0,0,1	80	0.00	0.00	4.02	2.333,0.5,0.5	0.00	0.00	2.25
1,1,1	150	0.06	Unstable	Unstable	2.333,0.5,0.5	0.00	4.42	4.42
1,1,1	150	—	—	—	2.333,0.7,0.7	0.00	4.25	4.25
1,1,1	150	—	—	—	2.333,0.85,0.85	0.00	4.32	4.32
1,1,1	150	—	—	—	2.333,0.95,0.95	0.00	5.57	5.57

Note: PI controller settings: Loop 1: $K_c = -0.600$, $\tau_1 = 0.20$; loop 2: $K_c = 5.767$, $\tau_1 = 6.23$; loop 3: $K_c = 5.767$, $\tau_1 = 6.23$.

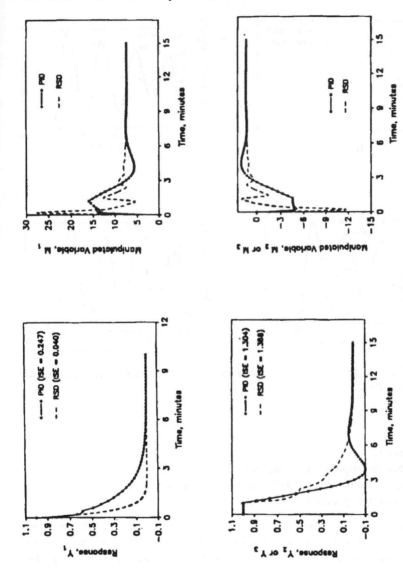

Figure 9.17 Regulatory performance for a unit step change in load affecting all loops in the absence of modeling errors.

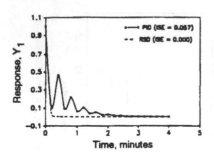

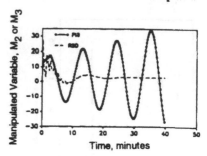

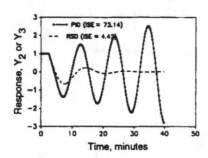

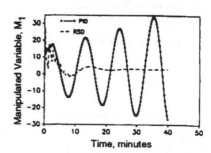

Figure 9.18 Regulatory performance in the presence of a 150% modeling error in each parameter of the 3×3 system.

where

O = output,
O_s = process output initial condition,
ΔI_j = input(manipulated variable) move, $I_j - I_{j-1}$,
k = kth sampling instant.

For sampling instants beyond $k + 1$, ΔI_{k+1-i} is zero and, therefore, the upper limit on the summation sign in Eq. (9.32) may be replaced by N, where N denotes the open-loop settling time in terms of the number of sampling intervals, giving

$$O_{k+1} = \sum_{i=1}^{N} a_i \Delta I_{k+1-i} + O_s \qquad (9.33)$$

Incorporating Effect of Disturbances

Equation (9.33) only takes into account the effect of input upon output. To account for the possible presence of disturbances and modeling errors, the predicted outputs may be corrected according to

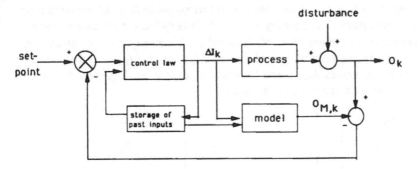

Figure 9.19 Block diagram of DMC system.

$$O_{k+1} = \sum_{i=1}^{N} a_i \Delta I_{k+1-i} + O_s + d_{k+1} \qquad (9.34)$$

On the basis of Eq. (9.34), the projected future outputs at sampling instants $k + L$ (where $L = 1, 2, \ldots$) may be computed according to

$$O_{k+L} = \sum_{i=1}^{L} a_i \Delta I_{k+L-i} + O_s + \sum_{i=L+1}^{N} a_i \Delta I_{k+L-i} + d_{k+L} \qquad (9.35)$$

For convenience, let us define

$$O_{k+L}^{\bullet} \equiv O_s + \sum_{i=L+1}^{N} a_i \Delta I_{k+L-i} \qquad (9.36)$$

Then, Eq. (9.35) may be expanded for $L = 1, 2, \ldots, (N + M)$ giving

$$O_{k+1} = O_{k+1}^{\bullet} + d_{k+1} + a_i \Delta I_k$$

$$O_{k+2} = O_{k+2}^{\bullet} + d_{k+2} + a_2 \Delta I_k + a_1 \Delta I_{k+1}$$

$$\vdots$$

$$O_{k+M} = O_{k+M}^{\bullet} + d_{k+M} + a_M \Delta I_k + a_{M-1} \Delta I_{k+1} + \cdots + a_1 \Delta I_{k+M-1} \qquad (9.37)$$

$$\vdots$$

$$O_{k+N} = O_{k+N}^{\bullet} + d_{k+N} + a_n \Delta I_k + a_{N-1} \Delta I_{k+1} + \cdots + a_{N-M+1} \Delta I_{k+M-1}$$

$$\vdots$$

$$O_{k+N+M} = O_{k+N+M}^{\bullet} + d_{k+N+M} + a_N \Delta I_k + a_N \Delta I_{k+1} + \cdots + a_N \Delta I_{k+M-1}$$

Note that the output projections have been carried out to $N + M$ sampling intervals into the future. The quantity $P = N + M$ is called the *prediction horizon*. The parameter M (where $M < N$) is called the *horizon of control calculations*. It represents the number of control moves, ΔI_k through ΔI_{k+M-1}, that are to be calculated; the moves ΔI_{k+M} and beyond are set to zero. Equation (9.37) may be written in a compact matrix form as

$$
\begin{bmatrix} O_{k+1} \\ O_{k+2} \\ \vdots \\ O_{k+N+M} \end{bmatrix} = \begin{bmatrix} O^*_{k+1} + d_{k+1} \\ O^*_{k+2} + d_{k+2} \\ \vdots \\ O^*_{k+N+M} + d_{k+N+M} \end{bmatrix}
$$

$$
+ \begin{bmatrix} a_1 & & & \\ a_2 & a_1 & \cdots & \\ \vdots & & & \\ a_M & a_{M-1} & \cdots & a_1 \\ \vdots & & & \\ a_N & a_{N-1} & \cdots & a_{N-M+1} \\ a_N & a_N & \cdots & a_N \end{bmatrix} \begin{bmatrix} \Delta I_k \\ \Delta I_{k+1} \\ \vdots \\ \Delta I_{k+M-1} \\ 0 \\ \vdots \\ 0 \end{bmatrix} \qquad (9.38)
$$

The second matrix on the right-hand side, which we denote by the symbol **A**, is called the *dynamic matrix* of the process. The process identification procedure described in Chapter 7 may be used to determine the elements of **A** (see also [17]). Note that the projected outputs $O_{k+N+1}, \ldots, O_{k+N+M}$ depend only on the inputs, $\Delta I_k, \ldots, \Delta I_{k+M-1}$.

The computation of the projected outputs by Eq. (9.38) requires the knowledge of the future values of the disturbance. Because such knowledge is unavailable, the best one can do is to set each of the d's equal to d_k. The expression for d_k may be obtained from Eq. (9.34) as follows:

$$
O_k = \sum_{i=1}^{N} a_i \Delta I_{k-i} + O_s + d_k \qquad (9.39)
$$

or

$$
O_k = O^*_k + d_k
$$

The process output O_k is available because it is measured. Therefore, d_k can be calculated according to

$$d_k = O_k - O_k^*$$
(9.40)

With this modification, Eq. (9.38) may be written as

$$
\begin{bmatrix}
O_{k+1} \\
O_{k+2} \\
\vdots \\
O_{k+N+M}
\end{bmatrix}
=
\begin{bmatrix}
O_{k+1}^* + d_k \\
O_{k+2}^* + d_k \\
\vdots \\
O_{k+N+M}^* + d_k
\end{bmatrix}
$$

$$
+
\begin{bmatrix}
a_1 \\
a_2 & a_1 & \cdots \\
\vdots \\
a_M & a_{M-1} & \cdots & a_1 \\
\vdots \\
a_N & a_{N-1} & \cdots & a_{N-M+1} \\
a_N & a_N & \cdots & a_N
\end{bmatrix}
\begin{bmatrix}
\Delta I_k \\
\Delta I_{k+1} \\
\vdots \\
\Delta I_{k+M-1} \\
0 \\
\vdots \\
0
\end{bmatrix}
$$
(9.41)

Formulation of the Control Problem

The purpose of the control effort is to calculate a set of control moves, ΔI_k, ..., ΔI_{k+M-1}, such that the projected outputs given in Eq. (9.41) follow the setpoint. To continue, we define a vector E according to

$$
E =
\begin{bmatrix}
O_s - O_{k+1} \\
O_s - O_{k+2} \\
\vdots \\
O_s - O_{k+N+M}
\end{bmatrix}
$$
(9.42)

Substituting for O_{k+j} from Eq. (9.41) into Eq. (9.42) gives

$$E = \begin{bmatrix} O_s - O_{k+1}^* - d_k \\ O_s - O_{k+2}^* - d_k \\ \vdots \\ O_s - O_{k+N+M}^* - d_k \end{bmatrix} - A \Delta I_k \tag{9.43}$$

where

$$\Delta I_k = [\Delta I_k \quad \Delta I_{k+1} \quad \cdots \quad \Delta I_{k+M-1}]^T \tag{9.44}$$

Denoting the first vector on the right-hand side as e_{k+1}, Eq. (9.43) may be written as

$$E = e_{k+1} - A \Delta I_k \tag{9.45}$$

Ideally, E should be zero but a unique solution is not possible because Eq. (9.45) has more equations than unknowns, i.e., it is an overdetermined set. The procedure that is followed is to formulate a least squares problem and solve it. The optimization index is

$$J(\Delta I) = E^T E = (e_{k+1} - A \Delta I_k)^T (e_{k+1} - A \Delta I_k)$$

$$= (e_{k+1}^T - \Delta I_k^T A^T)(e_{k+1} - A \Delta I_k) \tag{9.46}$$

or

$$J(\Delta I) = e_{k+1}^T e_{k+1} + \Delta I_k^T A^T A \Delta I_k - \Delta I_k^T A^T e_{k+1} - e_{k+1}^T A \Delta I_k \tag{9.47}$$

Taking the derivative of the terms in Eq. (9.47) and setting it equal to zero gives

$$\frac{\delta J}{\delta \Delta I_k} = 2 A^T A \Delta I_k - 2 A^T e_{k+1} = 0 \tag{9.48}$$

or

$$\Delta I_k = (A^T A)^{-1} A^T e_{k+1} \tag{9.49}$$

The first element of ΔI_k, namely, ΔI_k, is applied to the process at sampling instant k. At each sampling instant beyond k, beginning with $k + 1$, new measurement of O is taken and Eq. (9.49) is repeatedly used to establish closed-loop control of the process.

Formulation of DMC for Multivariable Systems

The problem formulation for multivariable systems proceeds in exactly the same fashion. Thus, for an r-output s-input multivariable system, the corrected outputs may be projected according to

$$\mathbf{O}_{k+1} = \sum_{i=1}^{N} \mathbf{a}_i \Delta \mathbf{I}_{k+1-i} + \mathbf{O}_s + \mathbf{d}_{k+1} \tag{9.50}$$

Here, $\mathbf{O}_{k+1}$ is an r-dimensioned vector of projected outputs, $\mathbf{a}_i$ is an $r \times s$ matrix of unit step response coefficients at the ith interval, $\Delta \mathbf{I}_k$ is an s-dimensioned vector of control moves at a given sampling instant, $\mathbf{O}_s$ is the initial condition vector, and $\mathbf{d}_{k+1}$ is the disturbance vector. For $r = s = 1$, Eq. (9.50) reduces to Eq. (9.34).

The dynamic matrix $\mathbf{A}$ of the multivariable system is composed of blocks of step response coefficient matrices of dimension $(N + M) \times M$ relating the ith output to the jth input arranged according to

$$\mathbf{A} = \begin{bmatrix} A_{11} & A_{12} & \cdots & A_{1s} \\ A_{21} & A_{22} & \cdots & A_{2s} \\ \vdots & \vdots & \vdots & \vdots \\ A_{r1} & A_{r2} & \cdots & A_{rs} \end{bmatrix} \tag{9.51}$$

The matrix $\mathbf{A}_{ij}$ contains all the ij coefficients of matrices $\mathbf{a}_L$, $L = 1, 2, ..., M$ arranged as in Eq. (9.38).

The projection vector for multivariable systems is

$$\mathbf{e}_{k+1} = [\mathbf{e}_{1,k+1}^T \ \mathbf{e}_{2,k+1}^T \ \cdots \ \mathbf{e}_{r,k+1}^T]^T \tag{9.52}$$

and the corresponding vector of moves is

$$\Delta \mathbf{I}_k = [\Delta \mathbf{I}_{1,k}^T \ \Delta \mathbf{I}_{2,k}^T \ \cdots \ \Delta \mathbf{I}_{s,k}^T]^T \tag{9.53}$$

With these definitions, Eq. (9.49) is valid for multivariable systems also.

In practical applications, it is usually necessary to suppress (restrict) the amplitude of the input moves. Move suppression in DMC is accomplished by adding additional rows to the matrix $\mathbf{A}$ giving

$$\mathbf{E} = \begin{bmatrix} \mathbf{e}_{k+1} \\ 0 \end{bmatrix} - \begin{bmatrix} \mathbf{A} \\ \Lambda \end{bmatrix} \Delta \mathbf{I}_k \tag{9.54}$$

where for multivariable systems

$$\Lambda = \text{diag}(\lambda_1 \ \lambda_1 \ \cdots \ \lambda_1 \ \lambda_2 \ \lambda_2 \ \cdots \ \lambda_2 \ \lambda_s \ \lambda_s \ \cdots \ \lambda_s) \tag{9.54a}$$

In Equation (9.54a), $\lambda_i > 0$ is the suppression coefficient for the ith move.

It is also possible to achieve tighter control of some controlled variables relative to others. This is achieved by premultiplying the equations by a matrix of weights Γ giving

$$E = \begin{bmatrix} e_{k+1} \\ 0 \end{bmatrix} - \Gamma \begin{bmatrix} A \\ \Lambda \end{bmatrix} \Delta I_k \tag{9.55}$$

where for multivariable systems

$$\Gamma = \text{diag}(\gamma_1 \ \gamma_1 \ \cdots \ \gamma_1 \ \gamma_2 \ \gamma_2 \ \cdots \ \gamma_2 \ \gamma_r \ \gamma_r \ \cdots \ \gamma_r) \tag{9.55a}$$

Here, $\gamma_i > 0$ is the weight assigned to the ith controlled variable. With these expressions for the weights on inputs and outputs, the least squares optimization index becomes

$$J(\Delta I_k) = \frac{1}{2} [A\Delta I_k - e_{k+1}]^T \ \Gamma^T\Gamma[A\Delta I_k - e_{k+1}] + \frac{1}{2} \ \Delta I_k^T \Lambda^T \Lambda \Delta I_k \tag{9.56}$$

The least squares solution of Eq. (9.56) is

$$\Delta I_k = (A^T\Gamma^T\Gamma A + \Lambda^T\Lambda)^{-1} A^T\Gamma^T\Gamma \ e_{k+1} \tag{9.57}$$

QDMC Approach

In real-life applications, the moves computed by Eq. (9.57) may or may not be implementable owing to process constraints that are usually present. The following types of constraints are usually encountered:

1. Manipulated variable constraints, for example, valve saturation;
2. constraints on the controlled variables, i.e., the outputs must be contained within allowable limits;
3. constraints on associated variables. These variables are not controlled but, for operational reasons, they must be kept within specified bounds; and finally
4. control moves need to be kept within certain bounds expressed by the inequalities $\Delta I_{min} < \Delta I_k < \Delta I_{max}$

Constraints on the projection of these variables may be expressed mathematically as a system of inequalities given by [18]

$$C\Delta I_k > c_{k+1} \tag{9.58}$$

where the matrix C contains dynamic information about the constraints, whereas the vector c_{k+1} contains the projected deviations of the constrained variables from their respective limits.

Constrained optimization problems of this type are ideally suited for quadratic programming approach for solving them. The optimization index in this instance is

$$J(\Delta I_k) = \frac{1}{2} \Delta I_k^T H \Delta I_k - g_{k+1}^T \Delta I_k \tag{9.59}$$

s.t. $C \Delta I_k > c_{k+1}$

$$\Delta I_{min} < \Delta I_k < \Delta I_{max}$$

where

$$H = A^T \Gamma^T \Gamma A + \Lambda^T \Lambda \qquad \text{(the QP Hessian matrix)} \tag{9.59a}$$

and

$$g_{k+1} = A^T \Gamma^T \Gamma e_{k+1} \qquad \text{(the QP gradient vector)} \tag{9.59b}$$

The solution of Eq. (9.59) by a quadratic programming algorithm produces an optimal set of moves ΔI_k which satisfy the constraints. Use of a commercially available QP algorithm has been suggested.

Tuning Constants in DMC

In DMC, the prediction horizon P is set equal to $N + M$. In other versions, there is freedom to choose another value for P although it has been pointed out that for minimum phase systems, P has virtually no effect on performance as soon as it exceeds twice the systems order. In the absence of modeling errors, with P set equal to $N + M$, stable system performance can been achieved even for nonminimum phase systems by choosing a sufficiently small value of M. Larger values of M give more freedom to the controller in matching the projected outputs and the setpoint. DMC will then deliver tighter control albeit at the expense of larger moves. Thus, the user has to balance quality considerations and the cost of achieving that quality.

In QDMC, the quality is additionally influenced by the projection interval to be constrained; only a subset of P ($= N + M$) constraints are usually constrained, starting with the Lth projection where $L > 1$. This subset forms a constraint window of future sampling intervals over which QDMC will ensure that constraint violations will not occur.

9.5 Additional Applications to Polymerization Reactors

In this section, three applications of multivariable control to free radical polymerization will be presented. The first involves the multivariable control of the solution polymerization of methyl methacrylate in a single CSTR. This is the same system which was controlled with the CMBC controller in Section 9.2. In this application, a true multivariable controller with implicit decoupling is used. The second application involves the same solution polymerization of methyl methacrylate, but carried out in a semibatch reactor. The unique problems of batch polymerization control are illustrated herein. The last application is to the control of a single CSTR in which emulsion polymerization of methyl methacrylate is taking place. As was noted in Chapter 7, this system exhibits oscillatory behavior as well as multiplicity, as seen in solution polymerization. The stabilization of the potential oscillatory behavior as well as the implicit decoupling of the very strongly interactive control objectives make this applications interesting.

9.5.1   Control of a CSTR Solution Polymerization Reactor

This study evaluates the application of pole placement adaptive controllers to a highly nonlinear polymerization process [14,15]. The principal objective is to establish the feasibility of implementation of this popular adaptive control structure and to elucidate the practical techniques required for successful closed-loop operation. Although numerous application studies detail promising results from pole placement controllers, the particular polymerization process selected presents some significant and unique control problems.

The simulation model is that of the solution polymerization of methyl methacrylate in a CSTR as described in Section 3.4. As previously noted, this system exhibits strong nonlinearities and multiple steady states. These dissipative structures result from high-energy liberations and poor heat dissipation, as well as high solution viscosity. The unusual open-loop dynamics provide many interesting and unique control problems. Conventional controllers are not suitably equipped to handle the time-varying dynamics observed in the polymerization reactor. Hence, adaptive controllers offer an intriguing potential alternative for controlling such processes. For this system, it can be shown that steady-state multiplicity results when the solvent concentration is less than 60 vol.% [14]. The solvent concentration in this study is maintained at 30 vol.% under these conditions, stable and unstable steady states result. The adaptive control strategies are used to regulate the system at both the upper and lower steady states.

Controller Design

The controller employed in this study is the pole placement/deadtime compensator of Vogel and Edgar [19–21]. From the block diagram in Fig. 9.20, it may be see that

$$\frac{y(z)}{y_r(z)} = \frac{B(z)H(z)z^{-k}}{A(z)G(z) + B(z)F(z)z^{-k}} \tag{9.60}$$

$D(z)$ is selected as the desired response of the closed-loop transfer function:

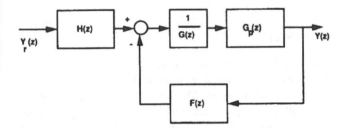

Figure 9.20 Block diagram of adaptive feedback control loop for the SISO system.

$$\frac{y(z)}{y_r(z)} = \frac{B(z)H(z)z^{-k}}{A(z)G(z) + B(z)F(z)z^{-k}} = D(z) = \frac{B_m(z)z^{-k}}{A_m(z)} \qquad (9.61)$$

where $D(z)$, $B_m(z)$, and $A_m(z)$ are polynomials in z. This formulation is similar to the approach taken in deriving the Dahlin algorithm. The numerator and denominator dynamics of the desired closed-loop transfer function can be arbitrarily selected. Vogel and Edgar select $A_m(z)$ to be a first-order response [16]:

$$A_m(z) = 1 - v_1 z^{-1} = 1 - e^{-T_s/T_{cl}} z^{-1} \qquad (9.62)$$

where T_s is the sampling period and T_{cl} is a tuning parameter. The numerator dynamics are selected to make the controller suitable for nonminimum phase plants. Thus, the zeroes of the desired closed-loop transfer function are equated with the process zeroes. The desired numerator, $B_m(z)$, is factored as

$$B_m(z) = B_{m1}B(z) \qquad (9.63)$$

where B_{m1} is a constant, yet unspecified. Likewise, the setpoint precompensator can be arbitrarily factored as

$$H(z) = H_1(z)B_{m1} \qquad (9.64)$$

Substituting Eqs. (9.63) and (9.64) into Eq. (9.61) yields

$$\frac{y(z)}{y_r(z)} = \frac{B_{m1}B(z)H_1(z)z^{-k}}{A(z)G(z) + B(z)F(z)z^{-k}} = \frac{B_{m1}B(z)z^{-k}}{A_m(z)} \qquad (9.65)$$

To incorporate setpoint tracking, the steady state closed-loop transfer function should approach unity. This is accomplished by setting $B_{m1} = A_m(1)/B(1)$.

From Eq. (9.65), it can be seen that the resulting characteristic equation for the closed-loop system is

$$A(z)G(z) + B(z)F(z)z^{-k} = A_m(z)H_1(z) \qquad (9.66)$$

The control engineer is free to select $H_1(z)$. Vogel and Edgar set $H_1(z) = A(z)$. If $H(z)$ and $F(z)$ polynomials are equated, the control law may be written

$$u(z) = \frac{H(z)}{G(z)[y_r(z) - y(z)]} \tag{9.67}$$

This controller is a type of multivariable Dahlin algorithm which incorporates deadtime compensation in much the same way as a Smith predictor [16]. Decoupling is implicit in the controller design. The form of the polynomials $A(z)$ and $B(z)$, as well as the length of the minimum time delay for each input/output pair, are fixed by the designer. The coefficients of $A(z)$ and $B(z)$ are determined by system identification as in Chapter 7. In this case, the identification is on-line via recursive least squares; it could also be done off-line.

Implementation

The polymerization process studied here is known analytically, making the selection of model orders somewhat trivial. The numerator and denominator polynomials are both selected to be third order. If the order is unknown, usually it is sufficient to assume a first-order process model; however, off-line identification may be warranted.

The model parameters can be updated on-line by a variety of estimation algorithms. In this study, recursive least squares is used. It is frequently applied in conjunction with adaptive control because of its rapid convergence and excellent model tracking. A key aspect of the estimation algorithm is establishing good initial model parameters and maintaining their integrity throughout the closed-loop run. A simple warm-up procedure is introduced in this study which effectively accomplishes plant start-up as well as initialization of the parameter estimator and controller. The warm-up procedure requires that an open-loop start-up policy be established. The manipulated variables are ramped from some initial values to the desired steady-state values. During the ramping cycle, the controller remains disengaged, while the estimator tracks the system. Because the measured states and manipulated variable continually change, sufficient information exists to adequately estimate the parameters. This procedure allows for a smooth transition from open to closed loop and helps to avoid undesirable start-up dynamics.

The pole location is the key design parameter for this controller. Pole location is set by adjusting the closed-loop time constants. The best pole location for the multivariable controller occurred at $z = 0.5$ for each input/output pair.

Results

The multivariable controller is employed to control the reactor temperature and monomer conversion by manipulating the jacket temperature and initiator feed concentration. System identification during start-up is employed as discussed. Fig. 9.21 charts the response of the system during start-up and under a load disturbance. The ramp warm-up procedure allows for an orderly transition from the open loop to the closed loop without introducing undesirable dynamics. These

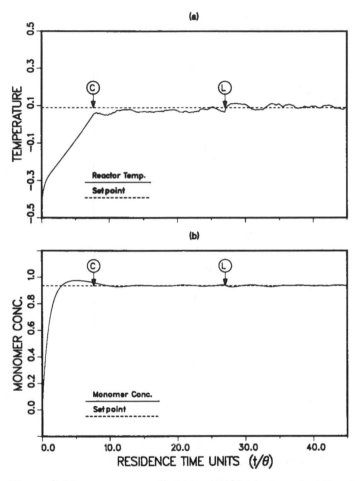

Figure 9.21 Performance of MIMO controller during a load disturbance. (a) Reactor temperature profile; (b) monomer concentration profile. (From Ref. 15.)

are especially encouraging results considering that the dynamic interaction between the input/output pairs are significant. Point L indicates when a step increase in the initiator efficiency factor, from 0.5 to 1.0, is injected into the plant. The multivariable regulator successfully maintains the desired setpoints for reactor temperature and monomer concentration, despite this extremely severe disturbance. A slight overshoot is observed in the temperature profile following the introduction of the load. The controller quickly damps this behavior without producing any detrimental transience. The controller outputs are well-behaved, requiring no excessive control action.

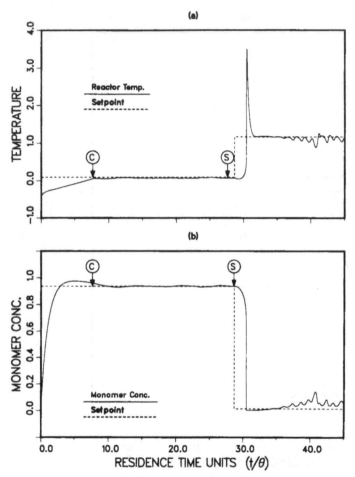

Figure 9.22 Performance of implicit MIMO controller during a setpoint change. (a) Reactor temperature profile; (b) monomer concentration profile. (From Ref. 15.)

Setpoint changes were also made. It is desired to move the process from a low monomer conversion steady state to a high conversion point. The results of the servo problem are illustrated in Fig. 9.22. Once again, the ramp warm-up procedure is implemented prior to commissioning the controller at point C. Point S indicates when setpoint changes for temperature and monomer concentration are made. The controller easily establishes these new values. Because of the severe gel effect inherent in the solution polymerization of MMA (methyl methacrylate) having low solvent concentrations, rapid reaction rates are accompanied by a large temperature surge. This accounts for the temperature spike exhibited in Figure 9.22. It is unlikely that any control scheme could be expected to damp this behavior, which is characteristic of this particular system.

The structure of this controller does not allow it to be used to control a system at an open-loop unstable point. Therefore, setpoint changes to such points are not attempted.

Conclusions

This adaptive pole placement controller performed adequately when applied to a highly nonlinear polymerization reactor. Both regulatory and servo control can be achieved. Decoupling is implicit and quite effective. The essential feature of the pole placement adaptive controllers is the location of the closed-loop poles. This would appear to be the only adjustable tuning parameter necessary to consider. However, tuning the controller to achieve a desirable closed-loop performance can be accomplished by other, less obvious means. The selection of sampling interval, model order, and modifications to the estimator all combine to affect the overall performance of the control scheme. Of course, some of these are not easily adjusted "on-line"; therefore, it is imperative that the choices available in designing the controller be made carefully.

9.5.2   Control of a Semibatch Polymerization Reactor

The previous application was for a continuous polymerization reactor where regulation is of most importance. This section deals with the control of a semibatch polymerization in which the control objectives are somewhat different [22]. Control of a semibatch reactor involves driving the reaction from an initial state to a specified final state in some manner which is judged to be the "best" in terms of productivity or product quality. Nonlinear model-predictive control may be employed to drive the reaction along a trajectory which maximizes some predetermined objective functional. Such a procedure requires an accurate model of the process which is often not available. To deal with this lack of knowledge about the process, an on-line, linear, time series model has been utilized. A discrete model-predictive control algorithm is employed to calculate the optimal controls to be applied to the polymerizer. These controls are optimal for the current control interval only, subject to the constraints of the process model and objective function; no attempt is made to optimize the entire trajectory.

Polymerization System

This "myopic" algorithm has been applied, in simulation, to the free radical solution polymerization of methyl methacrylate in a semibatch reactor as discussed in Subsection 3.4.3. The reaction is carried out in a semibatch configuration. Monomer and solvent are charged to the reactor. A free radical initiator is added during the course of the polymerization.

Controller

An analytical model of polymerization reaction is very important in improving the control of the system; however, the kinetics of most polymerization reactions are not accurately known. To deal with this, an empirical model of the process dynamics has been employed. The reactor model used is a linear, time series model of the following discrete form:

$$\mathbf{X}(K + 1) = \phi(K)\mathbf{X}(K) + \beta(K)\mathbf{U}(K) \tag{9.68}$$

where $\mathbf{X}$ and $\mathbf{U}$ are vectors of deviation variables of states and controls, respectively. Since this reactor system is highly nonlinear and the model is linear, the model is updated at each sample interval using the method of recursive least squares. The control algorithm employed minimizes an objective function of the following form:

$$I = \mathbf{X}^T(K + 1)\mathbf{F}\mathbf{X}(K + 1) + \mathbf{U}^T(K)\mathbf{E}\mathbf{U}(K) \tag{9.69}$$

where $\mathbf{F}$ and $\mathbf{E}$ are weighting matrices which assign the relative importance of minimizing each component in $\mathbf{X}$ and $\mathbf{U}$. The solution to this discrete, single-stage minimization is given by Bryson and Ho [23] as

$$\mathbf{U}(K) = K(K)\mathbf{X}(K) \tag{9.70}$$

where

$$K(K) = -[\mathbf{E} + \beta^T(K)\mathbf{F}\beta(K)]^{-1}[\beta^T(K)\mathbf{F}\phi(K)] \tag{9.71}$$

This controller minimizes the objective function, subject to the constraints of the linear model, for the next interval only. No attempt is made to optimize the entire trajectory because the adaptive model is only a linear approximation of the actual process during the current time interval.

Control of Temperature and Conversion

Two distinct control schemes are considered here. The first case is the control of reactor temperature and monomer conversion. The goal is to regulate the reactor temperature at $70\,°C$ while driving the conversion to its target of 85%. It is assumed that both temperature and monomer conversion can be measured on-line. Temperature is routinely monitored, and monomer conversion may be determined from refractive index [24] or density [25] measurements. For this situation the state and control vectors become

$$\mathbf{X} = \begin{bmatrix} (0.85 - C) \\ \left(\dfrac{T_d}{T_0} - \dfrac{T}{T_0}\right) \end{bmatrix}, \quad \mathbf{U} = \begin{bmatrix} (-q/v) \\ \left(\dfrac{T_{cd}}{T_0} - \dfrac{T_c}{T_0}\right) \end{bmatrix} \tag{9.72}$$

and the weighting matrices are

$$
\mathbf{E} = \begin{bmatrix} e_1 & 0 \\ 0 & e_2 \end{bmatrix}, \qquad \mathbf{F} = \begin{bmatrix} f_1 & 0 \\ 0 & f_2 \end{bmatrix} \tag{9.73}
$$

Since the initiator cannot be removed from the reactor, the initial initiator concentration was reduced by 50%, from the batch case, allowing the controller to add an additional initiator as needed. This tends to reduce initiator consumption and, because excess initiator is detrimental to end-use properties, improves product quality. By adjusting the weighting matrix coefficients, the emphasis placed on minimizing the state or control variables deviations is changed. This allows tailoring of the controller to meet operating requirements. If it is desired to control the temperature tightly and minimize initiator use, this can be accomplished by increasing e_1 and f_2. Setting the coefficients to $e_1 = 100$, $e_2 = 1$, $f_1 = 0.1$, and $f_2 = 100$ results in the control shown in Fig. 9.23. This plot shows the temperature controlled near its target with no initiator addition to the reactor. This results in a slow increase in conversion to target. If rapid polymerization is the main objective, f_1 is increased while the other three weighting coefficients are kept low. Figure 9.24 shows the control result when $e_1 = 1$, $e_2 = 1$, $f_1 = 10$, and $f_2 = 1$. The conversion time is greatly reduced by adding the initiator and increasing the reactor temperature. Whereas the conversion has been brought to target quickly as desired, the molecular weight of the product is unacceptably low. Variations in temperature will cause significant changes in the MWD. In this case, there is no control over temperature resulting in low MW, whereas the previous case exhibited good temperature control which kept the molecular weight high. A more desirable situation is a compromise between these two cases. This is achieved by setting $e_1 = 1$, $e_2 = 0.4$, $f_1 = 1$, and $f_2 = 50$. The control results for this case are shown in Fig. 9.25. The reactor temperature is kept closer to target which results in an average molecular weight that is on target.

Control of Conversion and Molecular Weight

Controlling temperature gives only indirect control over product quality. Because increasing conversion is counterproductive to keeping average molecular weight at an acceptable level, the batch time cannot be fully optimized unless MWD is controlled directly. A logical improvement is to directly control monomer conversion and number average molecular weight with initiator feed rate and coolant jacket temperature. On-line determination of MWD is commercially available via the on-line gel permeation chromatography. Molecular weight information between discrete chromatograms may be supplied by open-loop prediction or by an extended Kalman filter operating on the current time series model with rapid conversion feedback. The state vector then becomes

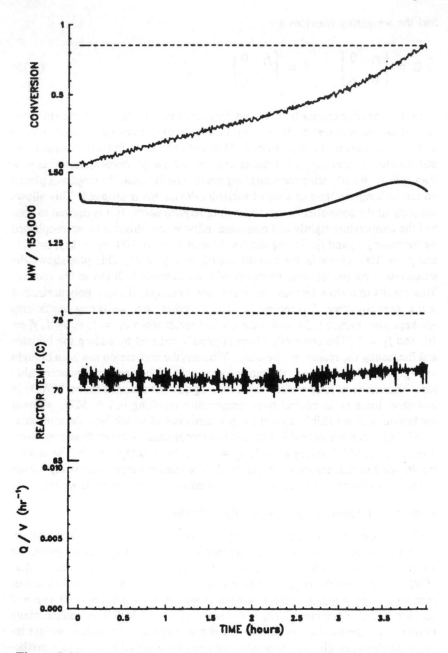

Figure 9.23   Adaptive predictive control of conversion and temperature: emphasis on temperature control. (From Ref. 22.)

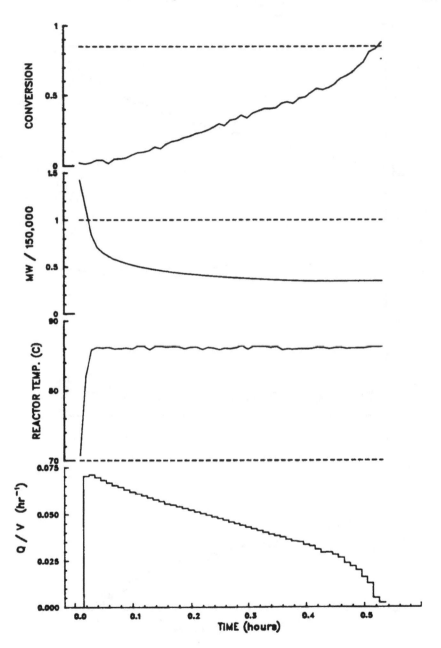

Figure 9.24 Adaptive predictive control of conversion and temperature: emphasis on conversion. (From Ref. 22.)

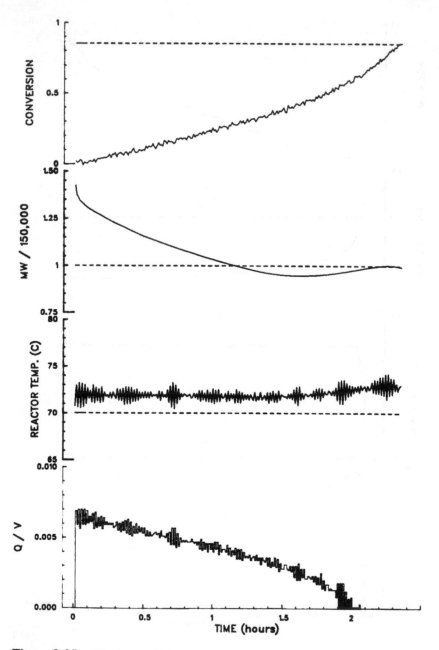

Figure 9.25 Adaptive predictive control of conversion and temperature: balanced emphasis. (From Ref. 20.)

$$\mathbf{X} = \left[\begin{array}{c} (0.85 - C) \\ \left(1 - \dfrac{MW}{MW_d} \right) \end{array} \right] \qquad (9.74)$$

The baseline case for this situation is a polymerization with PID temperature control as is shown in Fig. 9.26. An initiator concentration of 0.0218 g mol/L was chosen for the baseline because this is the amount which results in the correct average molecular weight. The batch time, molecular weight, polydispersity, and initiator use for this case are given in Table 9.3. Adaptive predictive control was applied to the system under the same conditions as the baseline case except that the initial initiator concentration was reduced to 0.0075 g mol/L. The emphasis in this case was placed on regulating average molecular weight, then optimizing conversion (reducing polymerization time) when possible. The results are shown in Fig. 9.27 and summarized in Table 9.3. This control results in a 2% decrease in reactor batch time and a 55% decrease in total initiator use. The results show tight control of molecular weight distribution. Note that the controller adds more initiator at the end of the batch to offset the gel effect. Because the average molecular weight produced is nearly constant, the resulting polydispersity is reduced to 1.626 from the baseline of 1.684.

One major advantage of adaptive predictive control is its ability to adapt to changing process conditions. To demonstrate this advantage, a process disturbance was applied to both the baseline and adaptive predictive control cases with all of the control parameters remaining the same. The disturbance selected was reduction in the initiator efficiency by 20%. This may be thought to represent the normal lot-to-lot variation in monomer and initiator reactivity. When this disturbance is applied to the baseline case, the batch time is increased by 10% and the molecular weight is increased by 9% (Table 9.3). When the same disturbance is applied to the adaptive predictive controller, more initiator is added to the reactor to compensate for the reduced initiator efficiency. Figure 9.28 shows that after an initial period of adjustment, the molecular weight remains on target throughout the batch cycle. Although 35% more initiator has been added to the reactor, there is little change in the molecular weight or polydispersity, as shown in Table 9.3.

Another process disturbance of considerable industrial concern is the level of residual inhibitor in the monomer. To keep MMA from polymerizing in storage, an inhibitor, such as hydroquinone, is added to the monomer in small quantities. Just before use, the inhibitor is removed from the monomer. Problems in the removal process may result in variations in the level of inhibitor remaining in the monomer charged to the reactor. This will result in variations in the polymerization rate. This occurs because the inhibitor reacts with free radicals as shown in Eq. (2.18). The concentration of free radicals is then given by Eq. (2.23). In this work, the assumption was made that k_{in} is 100 times k_i. The initial inhibitor concentration used in the simulation was 0.0075 g mol/L.

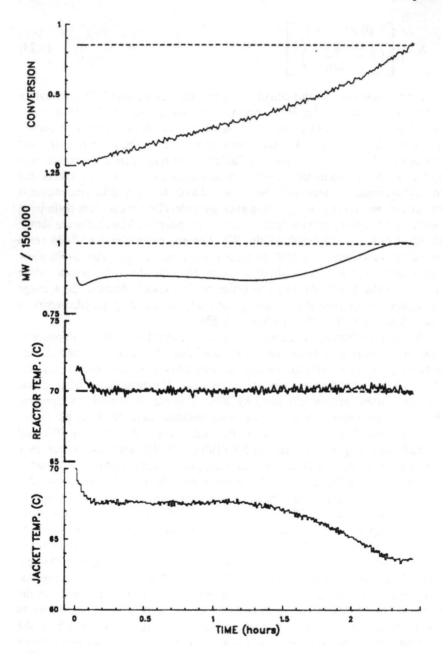

Figure 9.26 Baseline control case: PID temperature control; open-loop molecular weight control. (From Ref. 22.)

Table 9.3 Comparison of baseline and adaptive predictive control.

	No disturbance		Reduced initiator efficiency		Inhibited monomer	
	Baseline control	Adaptive predictive control	Baseline control	Adaptive predictive control	Baseline control	Adaptive predictive control
Batch time (hr)	2.44	2.38	2.72	2.34	2.76	2.47
Molecular weight ($m_n/m_{n\ desired}$)	0.999	0.989	1.085	0.991	1.071	1.015
Polydispersity	1.684	1.626	1.685	1.627	1.708	1.631
Initiator added during polymerization (g mol/L)	0	0.0048	0	0.0065	0	0.0069
Total initiator used (g mol/L)	0.0218	0.0099	0.0218	0.0108	0.0218	0.0110

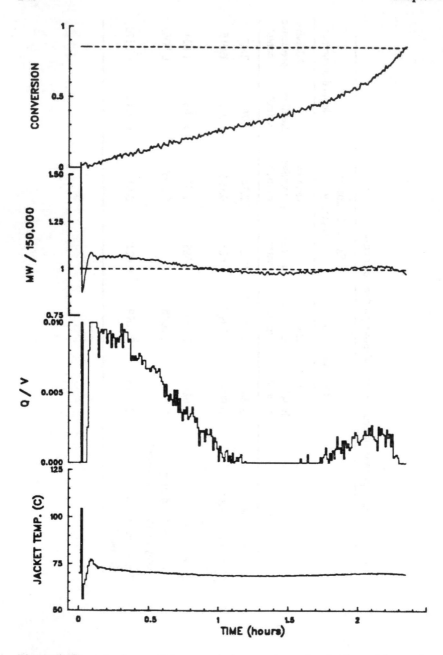

Figure 9.27 Adaptive predictive control of conversion and molecular weight: emphasis on molecular weight control. (From Ref. 22.)

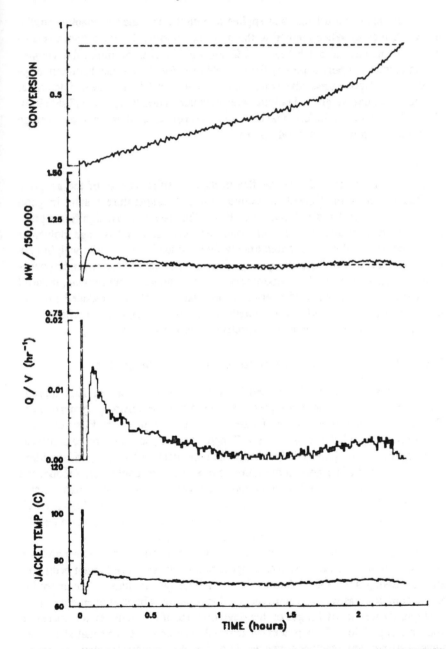

Figure 9.28 Adaptive optimizing control with poor initiator quality. (From Ref. 20.)

As before, this disturbance was applied to both the baseline and adaptive predictive control case. When applied to the baseline control, this disturbance results in a 13% increase in batch time, a 7% increase in average molecular weight, and an increase in polydispersity from 1.684 to 1.708. The same disturbance applied to the adaptive predictive control case results in a 4% increase in batch time, but the molecular weight and polydispersity remained nearly the same (Table 9.3). Figure 9.29 shows, as before, that after an initial period of adjustment, the average molecular weight was in good control.

Summary

Simulation results indicate the flexibility and effectiveness of the adaptive predictive algorithm. Control of conversion and temperature results in good temperature control and reduced batch time. The case of most significant practical importance is that of control of monomer conversion and average molecular weight because reduction of batch time without control of molecular weight may result in unacceptable polymer quality. Results indicate the following improvement over isothermal batch polymerization: reduced initiator consumption, reduced batch time, good control of average molecular weight, and reduced product polydispersity. The noted improvements are even more significant when variations in initiator and/or monomer reactivity are considered.

9.5.3   Control of Continuous Emulsion Polymerization

The previous applications have been for solution polymerization. This application involves a much more complex reaction system, emulsion polymerization. This adds a number of interesting features to the control problem. First, the reaction mechanism is heterogeneous (see Chapter 4), and, so, much more complex. This additional complexity is handled in this application by using on-line identification instead of a rigorous mechanistic model for the controller. Second, the effect of the control variables on monomer conversion and average particle size are strongly interactive; decoupling is a necessity. Finally, many continuous emulsion polymerization systems exhibit limit cycle behavior. Stabilizing these oscillations is a significant challenge for the control system.

Whereas batch and semibatch reactors predominate, continuous emulsion polymerization processes have found significant application due to their economic and processing advantages. Some of these advantages include lower operating and capital costs as well as improved product quality. The improvement of product quality frequently hinges on the development of efficient reactor operation and control policies. These processes generally comprise 2 to 15 equal-sized continuous stirred tank reactors connected in series. As previously noted, the operation of a train of CSTRs can be complicated by sustained oscillations in the conversion, number of particles, and all other related properties. Oscillations in latex

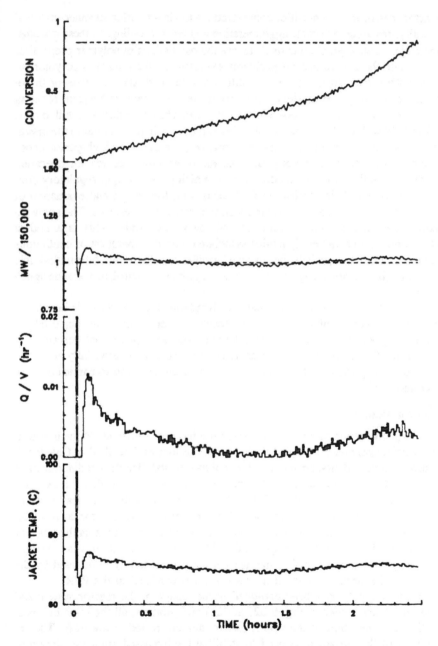

Figure 9.29 Adaptive predictive control with inhibited monomer. (From Ref. 22.)

reactors can produce emulsifier concentrations too low to offer adequate colloidal stability, resulting in particle agglomeration and reactor fouling. In these systems, the major control problem is to eliminate the oscillations in polymer properties.

Despite the unique control problems exhibited by the emulsion polymerization system, most such reactors are still controlled with standard two- or three-term, constant-gain controllers. These controllers may require frequent retuning to maintain satisfactory closed-loop performance. The controllers also fail to cope effectively with the large and variable deadtimes which are inherent in the operation of a train of emulsion reactors. However, the need for high-performance polymer products is increasing and, therefore, an improved reactor operation method as well as a robust control system which provides tight regulatory control are needed. This application [26] focuses on the design and evaluation of a control system for continuous emulsion polymerization reactors. The study involves the redesign and optimization of the reactor train. This is aimed at eliminating the unwanted dynamics associated with the operation of continuous emulsion reactors, especially the sustained oscillations. Removal of the undesirable dynamics by reactor design will facilitate regulatory control of the redesigned system.

The control system must be capable of handling large and variable deadtimes as well as severe nonlinearities of the reactor system. The controller used for the redesigned reactor train is the adaptive pole–zero placement described in Subsection 9.5.2. The use of an adaptive control strategy is especially attractive for emulsion polymerization systems in which the reaction dynamics are not well-modeled.

Reactor Design

The phenomenon of continuous oscillation has been attributed to the intermittent particle generation coupled with a slow washout as described in Chapter 7. The application of advanced control techniques to stabilize the oscillations have met with limited success [27,28]. Greene et al. [29] showed that the use of a PFR as the first reactor will stabilize the system. All particle nucleation takes place in the PFR. Subsequent growth of the particles takes place in the downstream CSTRs. The segregation of particle nucleation and growth prevents the onset of oscillation. This is shown in Figs. 9.30 and 9.31, where oscillations in the emulsion polymerization of methyl methacrylate are successfully eliminated by the use of a PFR seeder reactor if all nucleation is confined to the PFR.

These concepts have been exploited in the design of the reactor system and control scheme shown in Fig. 9.32. In this configuration, all the initiator, most of the surfactant part of the water, and monomer are fed to the PFR. The remainder of the surfactant is used to stabilize the bypassed emulsion stream to the first CSTR. The initiator flow rate to the PFR is used to control monomer conversion. The bypass of water around the PFR and directly into the CSTR is

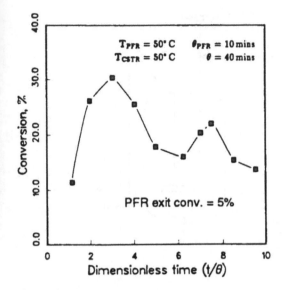

Figure 9.30 Emulsion polymerization in a PFR/CSTR reactor system (nucleation in the CSTR). (From Ref. 30.)

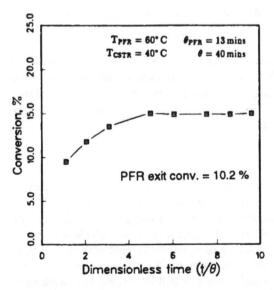

Figure 9.31 Emulsion polymerization in a PFR/CSTR reactor system (no nucleation in the CSTR). (From Ref. 30.)

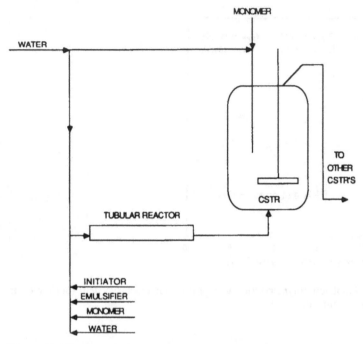

Figure 9.32 Emulsion polymerization reactor scheme. Monomer conversion and particle size control are via manipulation of initiator feed rate and water bypass. (From Ref. 26.)

used to control particle nucleation and hence, particle size. The variation of the aqueous flow split produces fluctuations in the surfactant concentration in the seeding reactor, resulting in changes in the particle number and, thereby, allowing for particle size control. It should be noted that the two controls are highly interactive and must be decoupled, explicitly or implicitly.

Reactor Simulation

A simulation model of the reactor system was developed. The kinetics of emulsion polymerization involves numerous elementary chemical and physical processes which operate simultaneously in both the organic particle phase and the aqueous phase. The dynamic behavior of a perfectly mixed CSTR is modeled by solving the unsteady-state material balance equations for each of the components of the reaction system. The dynamic equations for the tubular reactor were obtained by omitting the flow terms and treating each fluid element as a batch reactor with a reaction time equal to the residence time of the tubular reactor. Mass and energy balances were developed for initiator, radicals, inhibitors, surfactant, particle number, total polymer volume, and monomer and polymer. The dynamic model includes the following features and assumptions: density change in the reaction mixture as well as particle flocculation and breakup were considered negligible. Chain transfer reactions and radical desorption from particles were omitted. Inhibitor consumption by radicals was assumed to take place in the aqueous phase. Values for the physical constants and kinetic parameters were obtained from the open literature. The model includes correlations for the gel and glass effects.

The resulting eight coupled first-order differential equations describing the emulsion polymerization of MMA initiated by ammonium persulfate and stabilized by sodium lauryl sulfate (SLS) were numerically integrated. The modeling equations as well as more detail on the model development are available in the work of Temeng [30]. Simulation results show good agreement with laboratory polymerization results [26].

Control Strategy

The design of effective conventional controllers, in general, requires that the parameters of a model describing the dynamic behavior of the process be known. Even if an excellent process model is available, effective control over an extended period may not be attainable because of severe process nonlinearities, nonstationary characteristics, and large and variable deadtimes. Because controller parameters are frequently extracted by the linearization of a model in the vicinity of the operating point, the presence of severe nonlinearities will necessitate frequent retuning in response to changes in the desired operating conditions or detuning to balance stability and response over a range of operating conditions. In addition, deadtime compensators can fail to provide effective compensation because of their reliance on process models and the fact that deadtime usually varies during the operation of the plant. Satisfactory compensation can be obtained by tuning the controller for the largest expected delay, but variations in process conditions can render the control loop unresponsive.

Continuous emulsion polymerization reactors tend to be plagued by the process difficulties outlined above. The reactor dynamics are highly nonlinear and can exhibit multiple steady states during isothermal operation. Time-dependent dynamics frequently appear in the form of slow variations in the process parameters. Also, large and potentially varying deadtimes are inherent in the operation of a PFR system. For these reasons, an adaptive controller has been used. The controller chosen is that of Vogel and Edgar [19] which incorporates a linear time series model updated on-line via recursive least squares. This is the same controller used for the control of a continuous solution polymerization in Subsection 9.5.1. This multivariable controller allows for one on-line tuning parameter for each input–output pair and provides steady-state decoupling. In addition, the controller allows for different time delays among the input–output pairs.

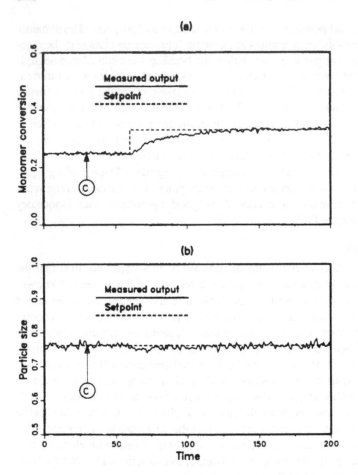

Figure 9.33 Controller performance during a setpoint change in conversion. (From Ref. 26.)

The controller was implemented for manipulation of a concentrated stream of initiator injected at the entrance to the PFR and the water flow split between the PFR and the CSTR to control monomer conversion and average particle size, respectively. The control of particle size has industrial significance because the viscosity and coating properties of the latex are dependent on the particle size.

Results

The multivariable adaptive control algorithm was applied to the simulation model described above. On-line measurement of monomer conversion (via density) and particle size (via light scattering) were assumed. White noise was added to the model inputs and outputs simulating actuator and sensor errors, respectively.

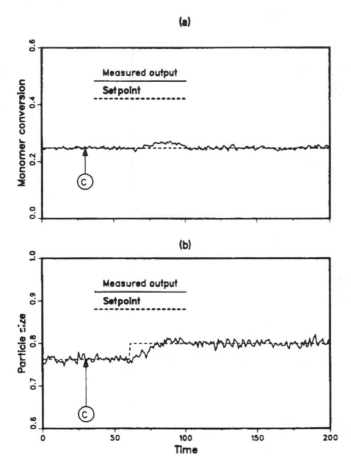

Figure 9.34 Controller performance during a setpoint change in particle size. (From Ref. 26.)

The servo problem was studied by making moderate changes in operating points for both states. Figure 9.33 shows the results for a setpoint change in conversion. The controller is implemented at $k = 30$ and at $k = 60$; it is desired to increase the conversion level from 0.247 to 0.33 while maintaining the average particle size constant. The controller is able to accomplish this task with no overshoot or oscillatory behavior despite significant variation in process gains. The adaptive controller was also used to make a setpoint change in the average particle size. The object is to move the scaled particle size from 0.76 to 0.80 while keeping the conversion level constant. Figure 9.34 shows that the controller is able to achieve this goal rapidly and with little overshoot.

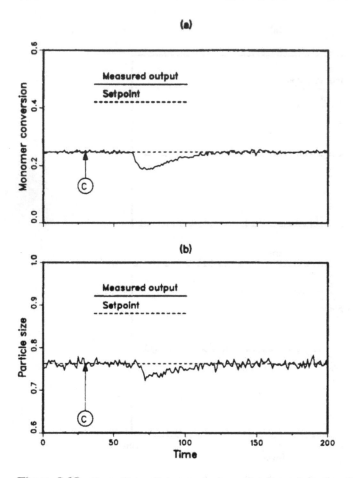

Figure 9.35 Controller performance during a disturbance in feed quality. (From Ref. 26.)

The ability of the controller to handle process disturbances was examined by simulating abrupt variations in feed quality. This was simulated by increasing the concentration of inhibitor in the monomer from 0 to 10 ppm. The variation in feed quality is a common problem in industrial practice and results from the deliberate addition of inhibitor to monomer to prevent premature polymerization. If monomer purification to remove inhibitor is not done (and it often is not in commercial operations), the switching of monomer feed tanks can produce undesirable and unexpected effects on the process outputs.

In the open loop, the increase in inhibitor concentration from 0 to 10 ppm causes a drop in the monomer conversion from 0.247 to 0.169. The particle size output also experiences a decline from 0.762 to 0.737. The reduction in polymerization rate is a direct result of the decreased initiator flux into the polymer particles, and the drop in particle size reflects the diminished conversion level. Figure 9.35 shows that the controller is able to eliminate the unacceptable consequences of the higher inhibitor concentration.

Conclusions

Synthesis of previous work in emulsion polymerization reactor dynamics revealed that conversion oscillations in a train of CSTRs can be eliminated by employing a tubular reactor as a seeder. Hence, the reactor configuration consisting of a CSTR preceded by a tubular prereactor was used to eliminate oscillatory behavior with the reactor design rather than with the control system. Simulation results demonstrate that adaptive control maintains a high level of closed-loop performance in both the servo and regulatory problems despite appreciable variations in the process dynamics and strong interactions between the manipulated variables. The adaptive design has the added advantage of not requiring a rigorous mathematical model for implementation.

References

1. P. B Deshpande (ed.), *Multivariable Process Control*, Instrument Society of America, Research Triangle Park, NC, 1989.
2. A. Uppal, W. H. Ray, and A. B. Poore, "On the Dynamic Behavior of Continuous Stirred Tank Reactors," *Chem. Eng. Sci. 29*, 967 (1974).
3. G. Mijares, "A New Criterion for the Pairing of Control and Manipulated Variables," *AIChE J.*, *32*, 1439–1449 (1986).
4. N. Jensen, D. G. Fisher, and S. L. Shah, "Interaction Analysis in Multivariable Control Systems," *AIChE J.*, *32*, 959–970 (1986).
5. E. Bristol, "On a New Measure of Interaction for Multivariable Process Control," *IEEE Trans. Automat. Control*, AC-11, 133–134 (1966).
6. A. Niederlinski, "A Heuristic Approach to the Design of Linear Multivariable Interacting Systems," *Automatica*, 7, 691 (1971).
7. T. J. McAvoy, *Interaction Analysis*, Instrument Society of America, Research Triangle Park, NC, 1983.
8. T. J. McAvoy, "Some Results on Dynamic Interaction Analysis of Complex Systems," paper No. 57d, AIChE National Meeting, New Orleans, LA, November 1981.
9. C. F. Moore, "Singular Value Decomposition," in *Multivariable Process Control* (P. B. Deshpande, ed.), Instrument Society of America, Research Triangle Park, NC, 1989.
10. H. Lau, L. Alvarez, and K. F. Jensen, "Synthesis of Control Structures by Singular Value Analysis: Dynamic Measures of Sensitivity and Interaction," *AIChE J.*, *31*, 427–439 (1985).

11. W. L. Luyben, "A Simple Method for Tuning SISO Controller in a Multivariable System," *Ind. Eng. Chem. Proc. Des. Dev.*, *25*, 456–460 (1986).

12. P. B. Deshpande and R. H. Ash, *Computer Process Control and Advanced Control Applications*, 2nd ed., Instrument Society of America, Research Triangle Park, NC, 1988.

13. P. R. Prasad, P. B. Deshpande, K. M. Kwalik, F. J. Schork, and K. W. Leffew, "Multiloop Control of a Polymerization Reactor," *Polym. Sci. Eng.*, *30*, 350–354 (1990).

14. K. M. Kwalik, M. S. thesis, Georgia Institute of Technology, Atlanta, GA, 1985.

15. K. M. Kwalik and F. J. Schork, "Adaptive Pole-Placement Control of a Continuous Polymerization Reactor," *Chem. Eng. Commun.*, *63*, 157–179 (1988).

16. N. V. Shukla, P. R. Krishnaswamy, and P. B. Deshpande, "Reference System Decoupling for Multivariable Control," *Ind. Eng. Chem. Res.*, 30, 662, 1990.

17. G. E. P. Box and G. Jenkins, *Time Series Analysis in Identification, Forecasting, and Control*, 2nd ed., Holden-Day, Oakland, CA, 1976.

18. C. E. Garcia and A. M. Morshedi, "Quadratic Programming Solution of Dynamic Matrix Control (QDMC)," *Chem. Eng. Commun.*, *46*, 73–87 (1986).

19. E. F. Vogel, "Adaptive Control of Chemical Processes with Variable Dead Time," Ph.D. thesis, University of Texas-Austin, 1982.

20. E. F. Vogel and T. F. Edgar, "Application of an Adaptive Pole-Zero Placement Controller to Chemical Processes with Variable Dead Time," *Proc. American Control. Conf.*, Arlington, VA, 1982, p. 536.

21. E. F. Vogel and T. F. Edgar, "An Adaptive Multivariable Pole-Zero Placement Controller for Chemical Processes with Variable Dead Time," presented at the AIChE Annual Meeting, Los Angeles, CA, November 1982.

22. W. E. Houston and F. J. Schork, "Adaptive Predictive Control of a Semibatch Polymerization Reactor," *Polym. Process Eng.*, *5*, 119–144 (1987).

23. A. E. Bryson and Y. C. Ho, *Applied Optimal Control*, Wiley, New York, 1975.

24. A. D. Schmidt and W. H. Ray, "The Dynamic Behavior of Continuous Polymerization Reactors – I. Isothermal Solution Polymerization in a CSTR," *Chem. Eng. Sci.*, *36*, 1401 (1981).

25. F. J. Schork and W. H. Ray," On-line Monitoring of Emulsion Polymerization Reactor Dynamics," in *Emulsion Polymers and Emulsion Polymerization* (D. R. Bassett and A. E. Hamielec, eds.), American Chemical Society, Washington, D.C., 1981.

26. K. O. Temeng and F. J. Schork, "Closed-Loop Control of a Seeded Continuous Emulsion Polymerization Reactor System," *Chem. Eng. Commun.*, *85*, 193–219 (1989).

27. K. W. Leffew and P. B. Deshpande, in *Emulsion Polymers and Emulsion Polymerization* (D. R. Basset and A. E. Hamielec, eds.), ACS Symposium Series, No. 165, American Chemical Society, Washington, D.C., 1981, p. 533.

28. C. Kiparissides, J. F. MaGregor, and A. E. Hamielec, *AIChE J.*, *27*, 13 (1981).

29. R. K. Greene, R. A. Gonzalez, and G. W. Poehlein, in *Emulsion Polymerization* (I. Piirma and J. L. Gardon, eds.), ACS Symposium Series No. 24, American Chemical Society, Washington, D.C., 1976, p. 341.

30. K. O. Temeng, Ph.D. thesis, Georgia Institute of Technology, Atlanta, GA, 1987.

10

Nonlinear Control Strategies for Polymerization Reactors

Nonlinearities are invariably present to some extent in industrial systems. In many cases, control system designs based on linearized behavior at specified operating conditions produce adequate results. However, it is widely recognized that properly designed nonlinear controllers are likely to yield much better results for nonlinear systems, especially as the extent of nonlinearities increase. Nonlinear systems dynamics and control is a hotly pursued topic in the current scientific literature.

Nonlinear systems exhibit a variety of interesting characteristics. One possible classification of nonlinear systems is as follows.

Type I Systems: These are single-input single-output systems. The unique feature here is that the system gain, deadtime, and/or time constant of the linearized model vary widely with the operating point. An example is pH control.

Type II Systems: These are two-dimensional systems involving two outputs (states) and one or more inputs. Typical examples are exothermic reaction systems where the outputs are temperature and conversion and the possible inputs are feed temperature, cooling rate, and catalyst (initiator) flow rate. These systems are potentially open-loop unstable. For certain choices of parameters and initial conditions, they exhibit multiple steady-states (output multiplicity). They could also exhibit input multiplicities, meaning that more than one value of an input give the same value of an output. The challenging problem here is to control the reactor at any of the steady states, especially the middle (of the three) unstable steady state in the face of disturbances and to shift the plant from one state to another.

Type III Systems: These are three- or higher-dimensional systems containing three or more outputs and one or more inputs. For certain choices of parameters and initial conditions, these systems can exhibit chaos. The challenging problem here is to control the chaotic system either at a single operating point or along one or more of its chaotic trajectories.

In all of the above cases, the presence of time delay further complicates the situation. There is preliminary evidence that nonlinear systems that otherwise do not exhibit chaos may do so in the presence of feedback control and/or time delays [1].

In polymerization systems, the difficulties are compounded by the fact that a large number of reactions are involved when initiation, propagation, chain transfer, and termination steps are considered. In spite of the intense interest of the scientific community, successful applications of multivariable nonlinear controller designs to polymerization reactors are not available in the open literature at present. What we present in this chapter is an introduction to the type of nonlinear control methodologies that are available. We hope the reader will conclude that these techniques do hold promise for polymerization systems. A review of nonlinear control methods is provided by Bequette [2]. For a text on nonlinear control the reader is referred to Isidori [3].

Several approaches to nonlinear control are presented in the following sections. The first approach is based on concepts from differential geometry. Given a nonlinear model of the plant relating the output Y to the manipulated variable U, the approach seeks to find a pseudo-manipulated variable, V, such that it is linearly related to the output Y. If such a transformation is possible, then any linear controller may be designed to obtain a linear closed-loop response from the nonlinear system. Global linearization is a method that follows this approach. Closely related to this method is one that is known by various names, namely, A Nonlinear System Controller, Reference System Control, and Generic Model Control.

The foregoing methods utilize conventional feedback control structures. The second approach we describe utilizes the internal model control structure. A simple method for deriving nonlinear IMC control laws will be described and some ideas on robustness will be presented. It will be seen that nonlinear IMC and the approaches based on differential geometry are closely related, and for certain systems the two will give similar results. We end the chapter with a discussion of controlling multiplicities and chaos in chemical reaction systems.

10.1 Approaches Based on Differential Geometry

10.1.1 Global Linearization

Given an SISO system having the form

$$\frac{dx}{dt} = f(x) + g(x)U \tag{10.1}$$

and

$$Y = h(x)$$

where x is a vector of states, Y is the measured output, and U is the manipulated variable, the global linearization [4] approach seeks to find a linear relationship between a new (pseudo) manipulated variable and the measured output. Lie algebra, a concept from differential geometry, is utilized to find such a transformation. The relationship is of the form

$$\frac{dx}{dt} = [f(x) + g(x)p(x)] + g(x)q(x)V \tag{10.2}$$

and

$$Y = h(x)$$

The relationship between V and U is given by the equation

$$U = \frac{V - \sum_{k=0}^{r} \beta_k L_f^k(h)}{\beta_r L_g L_f^{r-1} h(x)} \tag{10.3}$$

where L_g and L_f are the so-called Lie derivatives. The Lie derivative of a function $f(x)$ in the direction of the vector $h(x)$ is defined according to

$$L_f h(x) = \sum_{i=1}^{n} f_i(x) \frac{\delta h(x)}{\delta x_i} \tag{10.4}$$

It is also possible to differentiate $h(x)$ in the direction of $f(x)$ first, and then in the direction of $g(x)$, giving

$$L_g L_f h(x) = \sum_{i=1}^{n} g_i(x) \frac{\delta L_f h(x)}{\delta x_i} \tag{10.5}$$

Higher-order derivatives can also be written according to

$$L_f^k h(x) = L_f[L_f^{k-1} h(x)]$$ (10.6)

The term r in Eq. (10.3) is the relative order of the system representing the least positive number for which

$$L_g L_f^{r-1} h(x) \neq 0$$ (10.7)

The relative order represents the number of times Y must be differentiated with respect to t to recover the input U.

The transformation given in Eq. (10.3) yields a linear relationship between the output Y and the new input V according to

$$Y(s) = \frac{1}{\beta_0 + \beta_1 s + \cdots + \beta_r s^r} V(s)$$ (10.8)

Because the relationship between Y and V is linear, any linear controller can be used to control the output Y at setpoint. In particular, if a PI controller is used, the equation is

$$V(t) = \eta_1 e(t) + \eta_2 \int_0^t e(t) \, dt$$ (10.9)

where $e = (R - Y)$ and η_1 and η_2 are tuning constants. A block diagram of the globally linearized control system is shown in Fig. 10.1. It may be noted that, if the foregoing relationship between U and V is substituted into Eq. (10.3), a nonlinear control law containing PI terms will result from which the manipulated variable, U, may be directly computed.

10.1.2 Reference System Approach to Nonlinear Control

In this approach [5–7], it is desired to have a nonlinear system follow a desired linear reference trajectory. The commonly used reference system trajectories are

$$\frac{Y}{R} = \frac{1}{\tau s + 1} e^{\theta ds}$$ (10.10)

and

$$\frac{Y}{R} = \frac{2\zeta\tau s + 1}{\tau^2 s^2 + 2\zeta\tau s + 1} e^{\theta ds}$$ (10.11)

These specifications state that once the plant delay is over, the plant output will reach the setpoint according to a first- or a second-order trajectory. We shall work in the continuous time domain for illustrative purposes; the Laplace transform operator has been omitted for brevity. Let us select the second-order trajectory

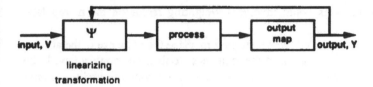

Figure 10.1 Global linearization concepts.

for the illustration of the method. Cross-multiplication of the terms in Eq. (10.11) followed by rearrangement gives

$$se^{\theta ds}\, Y(s) = \frac{2\zeta}{\tau}\, [R(s) - e^{\theta ds}\, Y(s)] + \frac{1}{\tau^2}\, [R(s) - e^{\theta ds}\, Y(s)]\, \frac{1}{s}$$

(10.12)

Equation (10.12) may be inverted giving

$$\frac{dY(t + \theta_d)}{dt}\, R = \frac{2\zeta}{\tau}\, e(t + \theta_d) + \frac{1}{\tau^2} \int_0^t e(t + \theta_d)\, dt$$

(10.13)

Equation (10.13) assumes that the designer is aware of the need to change the setpoint θ_d time units earlier. The symbol R on the left-hand side of Eq. (10.13) indicates that we wish the plant derivative to follow the right-hand side of the equation.

Now, suppose that the dynamic model of the plant is described by the nonlinear differential equation.

$$\frac{dY(t + \theta_d)}{dt} = f(Y, t, \lambda, U)$$

(10.14)

where the term λ represents model parameters and U is the manipulated variable. If it is required that the plant output follow the reference system exactly, then the right-hand sides of Eqs. (10.13) and (10.14) may be equated giving the nonlinear control law. Several comments on this design procedure are appropriate.

1.   The implementation of this strategy would require the availability of $Y(t + \theta_d)$, whereas in real-life applications only measurements of $Y(t)$ would be available. Therefore, $Y(t + \theta_d)$ would have to be predicted from a dynamic mathematical model of the plant. This prediction provides deadtime compensation.

2.   In some cases, the requirement that the plant output follow the reference system trajectory exactly may lead to an excessively active controller. In such cases, an optimization procedure may be invoked which would seek to minimize the difference between the actual and desired trajectories in the least squares sense.

The optimization index could involve a weighting factor on the manipulated variable as well.

3.   The terms ζ and τ allow the designer to shape the response in the absence of modeling errors. They are also the robustness tools in the control law. In the presence of modeling errors, they may be adjusted to maintain stability. Simulation work may be employed to get an idea about the region of stability because global stability may not be feasible.

4.   In the light of our discussion of global linearization, it should be clear that the specification of a linear relationship for the desired value of dY/dt is tantamount to assuming that the system is linearizable. If a linearizing transformation does not exist, then the real process output will deviate from the specified linear trajectory. In extreme situations, the strategy may not work at all.

10.2   Nonlinear Internal Model Control

If a mathematical model of the nonlinear plant can be developed, then a nonlinear IMC approach may be appropriate. A method for nonlinear IMC design based on operator theory is available [8]. In what follows, we describe a simple method of designing nonlinear IMC controllers [9]. To begin, let us briefly review how linear IMC controllers are designed. Consider for reasons of clarity a first-order process having the transfer function

$$G_p(s) = \frac{Y(s)}{U(s)} = \frac{1}{\tau s + 1} \tag{10.15}$$

The time domain equivalent of Eq. (10.15) is

$$\tau_p \frac{dY(t)}{dt} + Y(t) = U(t) \tag{10.16}$$

The IMC controller for this plant is the inverse of Eq. (10.16), i.e.,

$$G_c(s) = \frac{U(s)}{E(s)} = \tau_p s + 1 \tag{10.17}$$

Discounting the hardware realizability issues for the purpose of this discussion, we note that the choice of Eq. (10.17) for the controller makes the product of G_c and G_p unity, meaning that deadbeat responses will result for changes in setpoint. In the presence of modeling errors, a suitable filter may be used along with the controller to maintain robustness.

The time domain equivalent of Eq. (10.17) is

$$\tau_p \frac{dE}{dt} + E(t) = U(t) \tag{10.18}$$

where

$$E(t) = R(t) - [Y(t) - Y_M(t)]$$

Equation (10.18) is the IMC control law for the process whose transfer function is given in Eq. (10.15). Note that the variables in Eq. (10.18) are in deviation form.

A comparison of Eqs. (10.16) and (10.18) reveals that the IMC control law can be derived by substituting E for Y in Eq. (10.16). The procedure can be extended to higher-order systems and to nonlinear systems. The method of designing nonlinear IMC controllers for SISO and SIMO nonlinear systems then consists of the following steps.

1. Write an unsteady-state mathematical model of the process.
2. Solve the model equations for the manipulated variable.
3. Express the controlled variable(s) and its derivative(s) in terms of deviation variables.
4. Replace the deviation variables by E, representing the error input to the IMC controller to obtain the final form of the nonlinear IMC controller.

It may be pointed out that the steps involved in steps 2 could involve quadratic or higher-order terms, suggesting the possible presence of multiple steady states. The possibility that the closed-loop system may exhibit chaos for certain choices of initial conditions and parameter values cannot be ruled out. Presence of multiple steady states and chaos present certain difficulties for the exact nonlinear IMC controller which can be overcome as we will see.

The foregoing step-by-step procedure results in a controller that is based on the exact nonlinear inverse. Such designs can be extremely sensitive to modeling errors. Robustness can be incorporated to reduce the sensitivity in a finite parametric region according to the procedure that will be outlined shortly.

The design procedure will be illustrated by a number of examples. Each features a unique challenge as we will see. Unfortunately, application to polymerization reactions are only under investigation now and the results are not yet available.

10.2.1 Application to pH Control

In this case, the objective is to control the pH of an effluent from a CSTR by manipulating the flow rate of a base (NaOH). A strong-acid strong-base system is selected for the illustration of the method. The disturbance variable is taken to be the concentration of the incoming waste acid stream. It is desired to demonstrate servo and regulatory control capabilities of the nonlinear IMC approach.

The mathematical model of the system may be developed by writing unsteady-state mass balances around the process shown in Fig. 10.2. The resulting model is

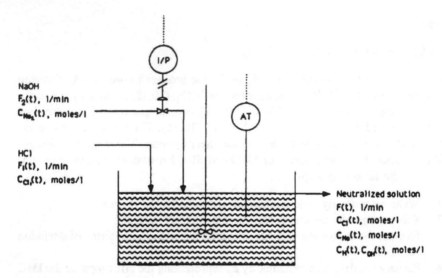

Figure 10.2   Schematic of pH system.

$$\frac{dY}{dt} = \frac{[Y + C_{cl_1} Y^2 - Y^3]P - [Y + C_{Na2} Y^2 - Y^3]Q}{1 + Y^2} \qquad (10.19)$$

where Y = dimensionless hydrogen ion concentration, $C_H/10^{-7}$, $P = F_1/60\,V$, $Q = F_2/60\,V$, and V = volume of the vessel. Solving Eq. (10.19) for the manipulated variable, Q, gives

$$Q = \frac{(1 + Y^2)z - [Y + C_{cl_1} Y^2 - Y^3]P}{Y + C_{Na2} Y^2 - Y^3} \qquad (10.20)$$

where z refers to the derivative dY/dt. Next, we introduce deviation variables in Eq. (10.20) according to

$$\bar{Y} = Y - Y_s \qquad (10.21)$$

Then Eq. (10.20) becomes

$$Q = \frac{[1 + (\bar{Y} + Y_s)^2]z - [(\bar{Y} + Y_s) + C_{cl_1}(\bar{Y} + Y_s)^2 - (\bar{Y} + Y_s)^3]P}{(\bar{Y} + Y_s) + C_{Na2}(\bar{Y} + Y_s)^2 - (\bar{Y} + Y_s)^3} \qquad (10.22)$$

Replacing Y by E in Eq. (10.22) gives

$$Q = \frac{[1 + (E + Y_s)^2]\,dE/dt - [(E + Y_s) + C_{cl_1}(E + Y_s)^2 - (E + Y_s)^3]P}{(E + Y_s) + C_{Na2}(E + Y_s)^2 - (E + Y_s)^3} \qquad (10.23)$$

Equation (10.23) is the IMC control law that is based on the exact nonlinear inverse. The presence of the derivative dE/dt in the controller equation can lead to the usual problems associated with the use of derivative terms in the controller equations. Note that E is defined according to

$$E = R - (Y - \hat{Y}) \tag{10.24}$$

The block diagram of the nonlinear IMC system for pH control is shown in Fig. 10.3. Studies indicate that the choice of exact nonlinear IMC leads to excessive variations in the manipulated variable, as expected, and an intolerance for modeling errors. Robustness may be built into the control system by incorporating a suitable filter. In this example, robustness may be introduced by incorporating an adapter which operates according to the equation

$$\frac{d\hat{\beta}}{dt} = \epsilon(R - Y) \tag{10.25}$$

where $\hat{\beta}$ = a suitable parameter in the model.

In Eq. (10.25) ϵ is a tuning parameter. The adapter serves as the filter. It attributes any differences between R and Y to an error in the model parameter, $\hat{\beta}$. No general guidelines on how to select which model parameter to adapt are available at this time. Simulation studies may be used to identify suitable model parameters for adaptation. The solution of Eq. (10.25) gives the value of the selected parameter for adaptation. Note that the selected model parameter will be present in the model as well as in the controller expression since the latter is the exact nonlinear inverse of the model. The tuning constant ϵ may be used to dampen the manipulated variable movements. Its value would have to be determined by simulation or by trial and error; no tuning guidelines are available at this time.

The choice of a pure integral for the adapter as depicted in Eq. (10.25) may not be unique although this form has been successfully used in a number of applications. A second application of this strategy is presented in the following section.

10.2.2 Application to CSTR-1

The system for this example is taken from the citation classic by Ray and associates [11]. It involves a jacketed continuous stirred tank reactor (CSTR) in which a first-order irreversible reaction $A \rightarrow B$ takes place. A coolant flowing through the jacket removes the heat of reaction to maintain the reactor outputs, conversion, and temperature, within specifications. The dimensionless mathematical model of the system may be developed by writing out the unsteady-state mass and energy balances and is given by

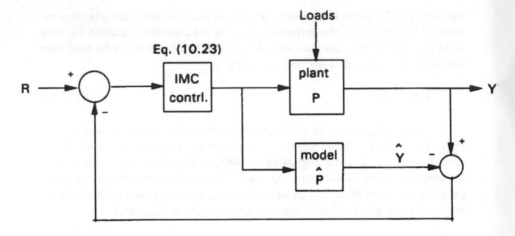

Figure 10.3   Block diagram of nonlinear IMC system.

$$\frac{dx_1}{dt} = -x_1 + Da(1 - x_1) \exp\left(\frac{x_2}{1 + (x_2/\gamma)}\right) - d_2 \qquad (10.26)$$

$$\frac{dx_2}{dt} = -x_2 + BDa(1 - x_1) \exp\left(\frac{x_2}{1 + (x_2/\gamma)}\right) - \beta(x_2 - x_{2c0})$$

$$+ \beta U_T + d_1 \qquad (10.27)$$

where x_1 represents dimensionless concentration and x_2 is dimensionless temperature. Ravi Kumar et al. [12] have applied the nonlinear IMC type approach for controlling the simulated system and the results will be briefly summarized in the following paragraphs. Ray and associates have shown that this system exhibits multiple steady states consisting of an upper and a lower stable steady state and a middle unstable steady state. In the absence of suitable feedback, tiny disturbances can drive the system away from the middle unstable steady state to one of the two stable steady states. The trajectory that the system may

Table 10.1 Characterization of stationary solutions for analysis.

Case no.	Process parameters	Stationary solution				
		U_T	x_1	x_2	Stability	Tag
1.	$D\alpha = 0.072$, $\beta = 0.3$	0.43	0.2028	1.3473	Stable	1a
	$\gamma = 20.0$, $B = 8.0$		0.3258	2.1041	Unstable	1b
	$x_{2c0} = 0.0$		0.7986	5.0140	Stable	1c
	$d_1 = 0.0$, $d_2 = 0.0$					
2.	Same as case 1	-0.2033	0.1286	0.7446	Stable	2a
			0.4996	3.0274	Unstable	2b
			0.7409	4.5126	Stable	2c
3.	$D\alpha = 0.32$, $\beta = 3.0$	0.0	0.6671	1.8344	Unstable	3a
	$\gamma \to \infty$, $B = 11.0$					
	$x_{2c0} = 0.0$					
	$d_1 = 0.0$, $d_2 = 0.0$					

follow is unpredictable and hence there is need for due caution. Pelligrini and Biardi [1] have shown that this system under PI control can exhibit chaos.

Application of the procedure outlined in the previous section leads to the following IMC control law

$$U_T = \beta_m^{-1}\{\dot{E}_2 + (E_2 + x_{2s}) - B[\dot{E}_1 + (E_1 + x_{1s}) + d_{2,m}]$$
$$+ \beta_m[(E_2 + x_{2s}) - x_{2c0}] - d_{1,m}\} \qquad (10.28)$$

The stationary solutions used in testing and analysis are shown in Table 10.1.

In the first test, it was attempted to move the system initially operating under stable conditions, corresponding to tag 1a, to an unstable state, tag 2b, in the multistationarity region. The results shown in Fig. 10.4 are indicative of the inability of nonlinear IMC to achieve the desired results. In the second test, the unstable state corresponding to tag 3a was selected. In this instance, the open-loop system shows oscillatory behavior as shown in Fig. 10.5. It was desired to regulate the system at the unstable point in the presence of a disturbance in $d_1 = 1.5$. Again, the nonlinear IMC approach does not work as may be deduced from the results shown in Fig. 10.6.

Ravi Kumar et al. [12] developed a new approach called RNCL (robust nonlinear control law) to overcome the problem. The block diagram of the RNCL strategy is shown in Fig. 10.7. RNCL combines nonlinear IMC and adaptive

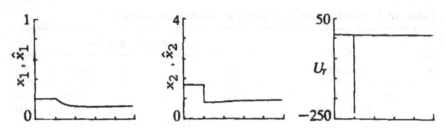

Figure 10.4   Servo responses of IMC (tag 1a — tag 2b, Table 10.1).

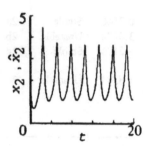

Figure 10.5   Open-loop response (unstable solution) corresponding to tag 3a.

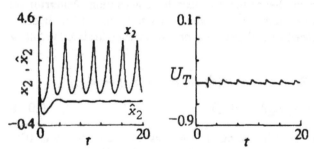

Figure 10.6   Regulatory performance of IMC ($d_1 = 1.5$).

control concepts whose purpose is to sever the dynamic attractive forces so that the system can be driven to the desired steady state instead of the state for which the system has the affinity. The system parameter β is used for adaptation purposes and a pure integral controller having the error ($x^{\text{set}} - x$) as the input used to update the selected parameter in the model and in the controller as shown in Fig. 10.7. The results shown in Figs. 10.8 and 10.9 indicate that RNCL works in cases where nonlinear IMC by itself does not.

The choice of a pure integral controller for the adapter may not be unique although this type of adapter has worked well in a number of case studies.

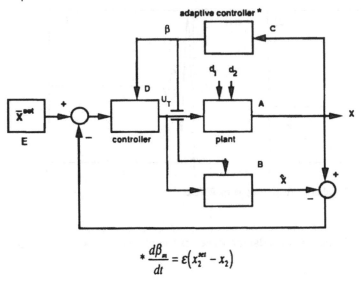

$$* \frac{d\beta_m}{dt} = \varepsilon \left(x_2^{set} - x_2 \right)$$

Figure 10.7 Schematic diagram showing implementation of RNCL.

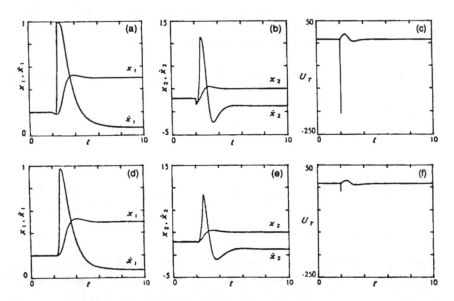

Figure 10.8 Servo responses with RNCL.

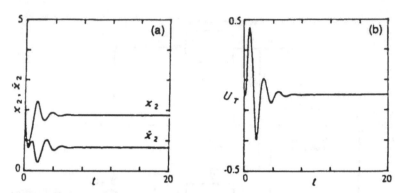

Figure 10.9 Regulatory responses with RNCL.

Studies also indicate that the adapter serves as the robustness tool for nonlinear IMC.

10.2.3 Application to CSTR-2

This example CSTR-2 [13] involves an irreversible consecutive first-order reaction $A \rightarrow B \rightarrow C$ in a continuous stirred tank reactor. Kahlert et al. [14] have shown that for certain choices of parameter values the system exhibits chaos. The goal in this instance is to control the system along one or more of its chaotic trajectories. The dimensionless mathematical model developed from unsteady-state mass and energy balances consists of the equations

$$\frac{dx_1}{dt} = 1 - x_1 - Dax_1 \exp\left(\frac{x_3}{1 + \epsilon x_3}\right) - d_2 \tag{10.29a}$$

$$\frac{dx_2}{dt} = -x_2 + Dax_1 \exp\left(\frac{x_3}{1 + \epsilon x_3}\right) - DaSx_2 \exp\left(\frac{\kappa x_3}{1 + \epsilon x_3}\right) - d_3$$
$$\tag{10.29b}$$

$$\frac{dx_3}{dt} = -x_3 + Dax_1 \exp\left(\frac{x_3}{1 + \epsilon x_3}\right) - DaB\alpha Sx_2 \exp\left(\frac{\kappa x_3}{1 + \epsilon x_3}\right)$$

$$- \beta(x_3 - x_{3c}) + \beta U_T + d_1 \tag{10.29c}$$

where x_1, x_2 are dimensionless concentrations of A and B, respectively; x_3 is dimensionless temperature; and U_T is the dimensionless coolant temperature. The parameters and their significance are outlined by Kahlert et al. [14]. The RNCL structure is useful in this case also, but the form of the adapter and its input are

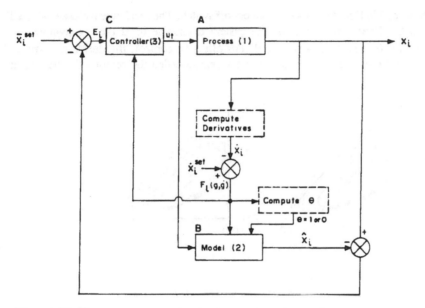

Figure 10.10 Schematic showing implementation of IMC for a chaotic process.

not the same. The required structure for the control of chaos is shown in Fig. 10.10. The justification for this structure for chaos control is as follows.

In chaotic systems, the specifications of the state variables alone do not completely characterize the state of the system and how it will evolve; the local derivatives must also be specified. The mathematical model of the system then takes on the form

$$\frac{d\hat{x}_1}{dt} = 1 - \hat{x}_1 - Da\hat{x}_1 \exp\left(\frac{\hat{x}_3}{1 + \epsilon\hat{x}_3}\right) - \hat{d}_2 + \theta F_1(g, \dot{g}) \quad (10.30a)$$

$$\frac{d\hat{x}_2}{dt} = -\hat{x}_2 + Da\hat{x}_1 \exp\left(\frac{\hat{x}_3}{1 + \epsilon\hat{x}_3}\right) - DaS\hat{x}_2 \exp\left(\frac{\kappa\hat{x}_3}{1 + \epsilon\hat{x}_3}\right)$$
$$- \hat{d}_3 + \theta F_2(g, \dot{g}) \quad (10.30b)$$

$$\frac{d\hat{x}_3}{dt} = -\hat{x}_3 + Da\hat{x}_1 \exp\left(\frac{\hat{x}_3}{1 + \epsilon\hat{x}_3}\right) - DaB\alpha S\hat{x}_2 \exp\left(\frac{\kappa\hat{x}_3}{1 + \epsilon\hat{x}_3}\right)$$
$$- \beta(x_3 - x_{3c}) + \beta U_T + \hat{d}_1 + \theta F_3(g, \dot{g}) \quad (10.30c)$$

The term θ in Eqs. (10.30a–c) is an on–off switch. The local eigenvalues evaluated from the linearized model of the system determine θ. If the eigenvalues are negative, θ is 1; otherwise it is 0. For effective control, the derivative corrections to the model must be made in the region called the region of entrainment

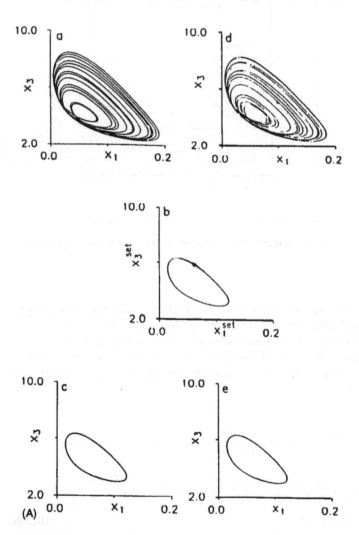

Figure 10.11 Phase plane plots of process outputs x_1, x_3 for servo control A: a,b,c: open loop, set, and controlled dynamics in the absence of load disturbance; d,b,e: open loop, set, and controlled dynamics in the presence of constant load disturbance. B: Controller action (u_t) in time for servo control (a: in the absence of load disturbance; b: in the presence of constant load disturbance).

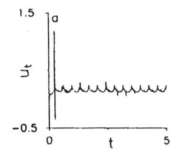

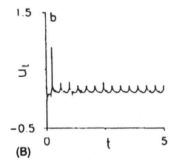

(B)

Figure 10.11 Continued

where the initial conditions $x_i(0)$ yield solutions of the model [(Eqs. (10.30a–c)] that converge to the specified chaotic setpoint trajectories. The presence of θ ensures that the system is always operating in the convergent region where, even in the presence of local perturbations, the system always tends toward the desired setpoint. The use of the model in Eqs. (10.30a–c) without the switch lead to the IMC control law

$$U_T = \frac{1}{\beta}[\dot{E}_3 + E_3 + x_{3s} - B\{1 - (E_1 + x_{1s}) - d_{2,m} + F_1(g, g) - \dot{E}_1\}$$

$$+ B\alpha[1 - (E_2 + x_{2s}) - (E_1 + x_{1s}) - d_{2,m} - d_{3,m} + F_1(g, g)$$

$$+ F_2(g, g) - \dot{E}_1 - \dot{E}_2] + \beta[E_3 + x_{3s} - x_{3c}] - d_{1,m} - F_3(g, g)]$$
$$\tag{10.31}$$

The performance of the control system was evaluated by simulation on a digital computer. The servo and regulatory responses shown in Figs. 10.11 and 10.12 show the excellent capability of the strategy for controlling the chaotic system.

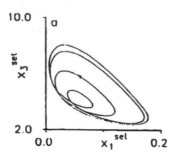

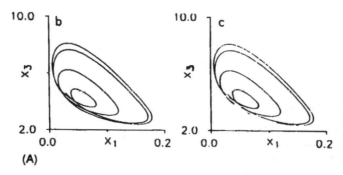

(A)

Figure 10.12 Phase plane plots of process outputs x_1, x_3 for regulatory control A: a,b: set and controlled dynamics in the absence of load disturbance; a,c: set and controlled dynamics in the presence of constant load disturbance. B: Controller action (u_r) in time for regulatory control (a: in the absence of load disturbance; b: in the presence of constant load disturbance).

This should be considered as a theoretical approach for now because the effect of time delays and model uncertainties have not been considered.

10.3   Applications to Polymerization Reactors

Almost all chemical process systems are nonlinear. Some are only slightly nonlinear, and so can be analyzed and controlled effectively with linear techniques. Others are so nonlinear that some form of nonlinear control may be desirable or even necessary. Among all the unit operations of chemical engineering, reactors tend to be some of the most nonlinear due to the Arrhenius functionality of variations in the reaction rate constants with temperature and higher-order overall reaction rates. Among reactors, polymerization reactors tend to be some of the most nonlinear due to the heat of reaction (highly exothermic) and the complex reaction networks. Because, in addition, polymer property specifications may be

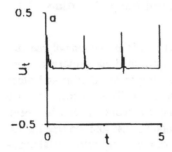

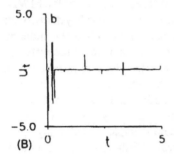

Figure 10.12 Continued

stringent, nonlinear control of polymerization reactors is often justified. Some nonlinear control is used routinely in polymerization facilities. Examples of these are tune-on-demand regulators and regulators for batch reactors in which the gain of the controller is varied in a preprogrammed manner as the reaction proceeds. Other, more complex schemes just are now being applied. These are model-based control methods of the type discussed above.

In this section we will discuss two applications of nonlinear control to polymerization reactors. The first is the application of a nonlinear regulator to a continuous solution polymerization reactor. Because these reactors are known to be unstable over much of the operating space (see Chapter 3), the emphasis here is on stabilization of the reactor at an unstable steady state. In the second application, a nonlinear model-predictive control algorithm is applied to a semibatch solution polymerization reactor. Here the emphasis is on nonlinear regulation, but also on using the nonlinear predictive capability of the controller to reach a desired set of polymer properties at the end of the polymerization. It is this "targeting" capability that makes this controller interesting.

10.3.1 Nonlinear Reference Control of Continuous Solution Polymerization

In this work, the problem of stabilization and regulation of the solution polymerization of methyl methacrylate in a single CSTR is approached via a reference controller [15]. This is a multivariable, nonlinear analog to the Dahlin and other response-specification methods discussed in Chapter 11. This control structure was proposed by Boye and Brogan [16]. Subsequent work on this has been presented by Bartusiak et al. [17]. Recall from Chapter 3 that solution free radical polymerization in a CSTR can be described by seven ordinary differential equations describing material balances over monomer, initiator, and solvent, an enthalpy balance, and material balances over the first three moments of the MWD. Recall also that this seven-state system can be broken into two subsystems, with the material balances on monomer, initiator, and solvent as well as the enthalpy balance being grouped into Subset I. Subset II, then contains the three differential equations for the first three moments of the MWD. The significance of the division is that whereas Subsystem II is affected by Subsystem I, the converse is not true. The strategy in this application is to stabilize and regulate Subsystem I at specified operating conditions which will produce a polymer of the correct MWD. Thus, the closed-loop control of Subsystem I is implemented. Only open-loop (through the setpoints of Subsystem I) control of Subsystem II is obtained. This is a common approach because on-line determination of MWD is rarely available. In this application, the first three states (monomer concentration, temperature, and initiator concentration) are controlled while allowing solvent concentration to vary slightly. This will be done by manipulating monomer feed rate, cooling jacket temperature, and initiator feed rate. In this section, the control scheme will be developed, and then implemented in simulation, showing that global stabilization can be achieved in some instances in spite of the fact that some of the open-loop steady states are unstable. (See Chapter 3.)

Nonlinear Reference Model Control

Here the subset of the state vector containing the first three state variables (monomer, temperature, and initiator levels) of Subsystem 1 will be controlled. This may be written as

$$\dot{x} = f(x) + Gu \tag{10.32}$$

where the model equations are nonlinear in the states, but linear in the controls. Inspection of the system equations in Chapter 3 confirms that this is the case. A reference system for the states is then defined as

$$\dot{x}_{r1} = A_1(x_{r1} - x_{d1}) \tag{10.33}$$

The vector x_{d1} consists of the desired values of monomer concentration, temperature, and initiator concentration. A_1 is chosen to be of the form kI, where

I is the identity matrix and k is some scalar, $k > 0$. In this approach, the value of k specifies the speed of response of the system to a change in setpoint or load and is specified by the designer. Defining the error as $e_1 \equiv (x_1 - x_{r1})$, and combining this definition with Eqs. (10.32) and (10.2), the dynamics of the error are given by

$$\dot{e}_1 = A_1 e_1 + [f_1(x) - A_1(x_1 - x_{d1}) + G_1 u] \tag{10.34}$$

The important point here is that Eq. (10.34) can be solved *exactly* for the control u required to make the error system e_1 asymptotically stable. By equating the term in the square bracket in Eq. (10.34) to zero, we have that

$$u = -G_1^{-1} [f_1(x) - A_1(x_1 - x_{d1})] \tag{10.35}$$

The design matrix A_1 controls the exponential decay of the error and, hence, the aggressiveness of control. For a value of A_1 which does not saturate the controls, the approach of the controlled variables to a new setpoint will be linear and first order despite the fact that the controlled system is nonlinear.

Simulation Results

Figure 10.13 shows the servo and stabilizing capabilities of the nonlinear reference controller. For this simulation, the reactor residence time is fixed at 53.33 min and the reactor parameters are chosen such that there are three steady states possible. The solvent feed rate is set to give a volume fraction of 64% at the beginning of the simulation and left unchanged. Under these conditions, there are three possible steady states. As is expected, the intermediate equilibrium point (x_{e2}) is unstable. The value of k is 0.6. Starting from an arbitrary initial condition, the reference controller moves the system to each of the open-loop steady states, including the unstable steady state. Note also the first-order dynamics of the first three states (monomer concentration, temperature, and initiator concentration). This is characteristic of the reference controller design. Because solvent concentration is not controlled, it does not exhibit first-order dynamics. Note that when steady state is reach once again, the solvent concentration returns to its initial value. Figure 10.14 shows the manipulated variables. Note that none of the inputs is saturated at this value of k. If k is set too high, it is possible that one or more on the inputs will saturate and the first-order dynamics will no longer apply.

The nonlinear reference controller provides zero offset on setpoint change, even without explicit integral action. Unfortunately, the same is not true for disturbance rejection. Integral action may be incorporated into the controller to provide disturbance rejection capabilities. The addition of integral action to the controller is achieved by state augmentation. It has been shown [18] that for a controllable *linear* system to remain controllable under integral state feedback, it is necessary that

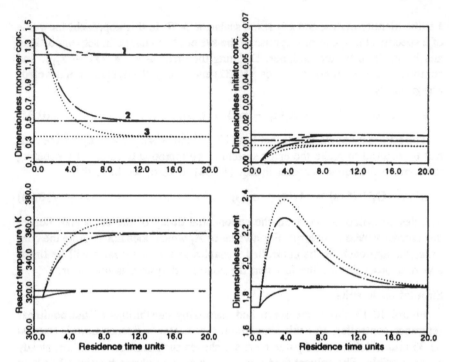

Figure 10.13 Nonlinear reference control servo control. Bringing the reactor from an arbitrary initial condition to the three open-loop steady states ($k = 0.6$). (1: upper steady state [stable], 2: intermediate steady state [unstable], 3: lower steady state [stable].) (From Ref. 19.)

$$n_1 \leq m \tag{10.36}$$

where n_1 is the number of state variables being fed back under integral action and m is the number of manipulated variables. The integral action is adjoined to the proportional reference controller input $\mathbf{u}$ in the following manner:

$$\mathbf{u} = -\mathbf{G}_1^{-1}[\mathbf{f}_1(\mathbf{x}) - \mathbf{A}_1(\mathbf{x} - \mathbf{x}_d)] - \mathbf{\Gamma}_i \int_0^t (\mathbf{x} - \mathbf{x}_d) \, dt \tag{10.37}$$

where $\mathbf{\Gamma}_i$ can be viewed as a diagonal matrix of non-negative integral gains.

In Fig. 10.15, a 10% decrease in the heat transfer coefficient (β) is introduced. This disturbance enters the temperature loop directly and no other loop, and it is assumed to be unmeasured. Hence, the controls are computed based on the *old* value of β. The controller parameters are $k = 2$ and an integral gain of 1 on the temperature loop (x_2). Because of the nature of the reference controller, one would expect the monomer (x_1) and initiator (x_3) loops to eventually return to the desired operating point after some dynamics *even without integral action*.

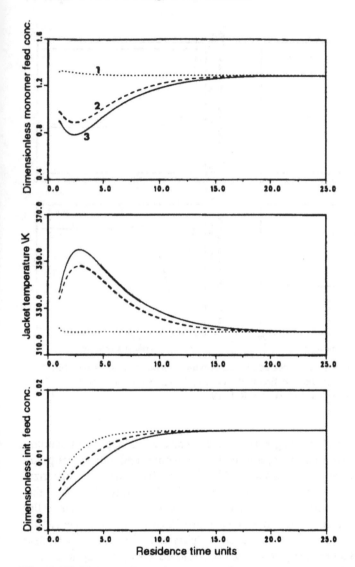

Figure 10.14 Nonlinear reference control manipulated variables. Manipulated variables corresponding to Figure 10.13. (From Ref. 19.)

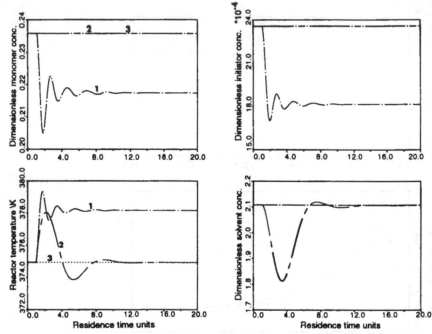

Figure 10.15 Nonlinear reference control disturbance rejection. Disturbance rejection: 10% reduction in heat transfer coefficient (Curves 1, 2, and 3 are the open loop, closed loop, and setpoint, respectively.) (From Ref. 19.)

Furthermore, because of the integral action in the temperature loop, at equilibrium, no offset exists in this loop. Also, note that the monomer and initiator loops are virtually unaffected by this disturbance. This follows from the exact linearization property of the *proportional reference controller*. Note from Fig. 10.27 that ultimately, there is no offset in the solvent loop (x_4). In effect, the controller has perfectly rejected the disturbance in the entire Subsystem 1 (and hence, Subsystem 2, in the absence of a disturbance to Subsystem II directly). It is clear that before the disturbance enters the temperature loop, the system is at equilibrium. The offset free property of the temperature loop after the disturbance in β is introduced is guaranteed by the integral action. Consequently, after the disturbance propagates, as far as the monomer and initiator loops are concerned, nothing has changed in the temperature loop when equilibrium is again attained. In effect, at equilibrium both the manipulated variables $(x_{1f}$ and $x_{3f})$ return to the values they took on before the disturbance entered the process. Hence, due to material balance considerations, x_{4f} is unaltered at equilibrium, in which case, there will ultimately be no offset in x_4.

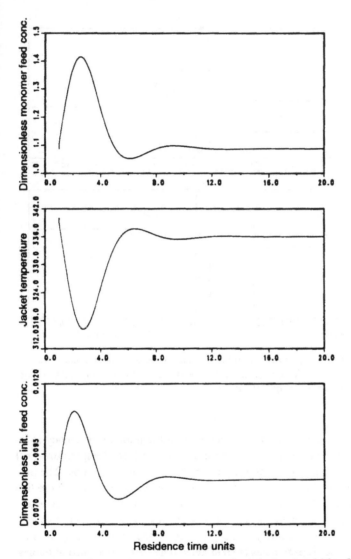

Figure 10.16 Nonlinear reference control manipulated variables. Manipulated variables corresponding to Fig. 10.15. (From Ref. 19.)

To see this more clearly, the input corresponding to Fig. 10.15 is presented in Fig. 10.16. Observe that after some transient dynamics, the monomer and initiator feed concentrations (x_{1f} and x_{3f}) return to their initial values. Of course, the new jacket temperature (T_c or x_{2c}) is correspondingly lower.

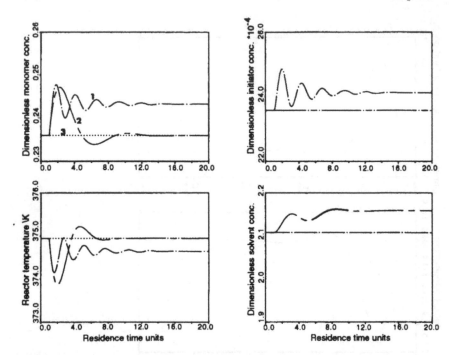

Figure 10.17 Nonlinear reference control disturbance rejection. Disturbance rejection: 10% reduction in initiator efficiency (Curves 1, 2, and 3 are the open loop, closed loop, and setpoint, respectively.) (From Ref. 19.)

In Fig. 10.17, a disturbance of 10% decrease in the initiator efficiency is introduced. This disturbance enters only the monomer (x_1) and temperature (x_2) loops directly. Now, integral action is incorporated in both these loops. The proportional and integral gains are set to values of 1 and 2, respectively. Note that there are two loops with integral action. Observe that the disturbance is effectively rejected in the controlled subset of the state. Note the offset in the solvent loop in Fig. 10.17. This ultimately leads to offsets in the MWD (Subsystem 2). In this case, disturbance rejection has been affected only in the controlled variables, and not in the solvent concentration.

Summary

In this investigation, the applicability of nonlinear reference control schemes for solution polymerization reactors has been demonstrated. The controller is easy to implement, provides stabilization of the unstable steady states, and, for some disturbances, perfect compensation is achieved for the four-state system with three controls. For other disturbances, compensation is achieved for the three controlled states only. If solvent concentration must be controlled as well for MWD considerations, a fourth manipulated variable must be identified. The nonlinear reference controller requires the availability of the state variables of Subsystem 1. In practice, these are not always available and must be estimated. Adebekun and Schork [19] have coupled the reference controller with an extended Kalman filter which estimates the states of Subsystem II from measurements of reactor temperature and monomer concentration (via densitometry).

10.3.2 Nonlinear Model-Predictive Control of a Semibatch Polymerization Reactor

In this section a nonlinear model-predictive control method is developed and applied to a semibatch solution free radical polymerization reactor [20]. The algorithm uses an explicit nonlinear process model and the basic elements of the classical dynamic matrix control (DMC). Update of the DMC model by a disturbance vector which accounts for the effect of nonlinearities in the prediction horizon is the key feature of the method. Its application to semibatch polymerization control is significant because this is one of the few algorithms which allows one to tackle the targeting problem common to all semibatch polymerizations. Here the emphasis is on nonlinear regulation, but also on using the nonlinear predictive capability of the controller to reach a desired set of polymer properties at the end of the polymerization. It is this targeting capability that makes this controller interesting in polymerization control.

Model-predictive control (MPC) is now widely recognized as a powerful methodology to address industrially important control problems. This is supported by many reported industrial applications and academic studies [21]. The general strategy of MPC algorithms is to utilize a model to predict the output into the future and minimize the difference between this predicted output and the desired one by computing the appropriate control actions. In particular, one MPC technique, dynamic matrix control DMC, has been studied very extensively [22].

Nonlinear Model-Predictive Control

Most MPC techniques, including DMC, are based on linear models and are thus not very well suited for the control of nonlinear systems. Because of this, there have been numerous efforts to extend MPC techniques for the control of nonlinear systems [23–25]. In this section an MPC algorithm is presented which uses the main structure of DMC, and yet explicitly accounts for nonlinearities in the process model.

Linear DMC The extended DMC algorithm developed here relies heavily on the standard, linear DMC control law discussed earlier. To establish a point of departure, we will review the major features of DMC. The major elements of DMC are: (a) the model, (b) estimation of the disturbance and projection into the future, and (c) computation of the control inputs.

Modeling Consider the single-input, single-output case without any loss of generality. The model used is a discrete "step response model" of the form

$$y(k) = \sum_{i=1}^{N} a_i \Delta u(k - i) + a_N u(k - N - 1) + d(k) \qquad (10.38)$$

where

k is the sampling time,
u is the input,
a_i are the step response coefficients,
d is the unmodeled or disturbance effects on the output,
N is the number of step response coefficients needed to adequately describe the process dynamics,
$\Delta u(k)$ is the change in the input defined as $u(k) - u(k - 1)$.

Estimation of the Disturbance Effects The current value of the disturbance effects can be estimated by subtracting the effects of past inputs on output from the current measurement of the output, i.e.,

$$d(k) = y^{\text{meas}}(k) - \sum_{i=1}^{N} a_i \Delta u(k - i) - a_N u(k - N - 1) \qquad (10.39)$$

Prediction of Outputs into the Future In the sequel, capital and lowercase boldface letters will denote matrices and vectors, respectively. The linear estimates of the future outputs are given by

$$
\underbrace{\begin{bmatrix} y(k+1) \\ y(k+2) \\ y(k+3) \\ \vdots \\ y(k+M) \\ \vdots \\ y(k+P) \end{bmatrix}}_{\mathbf{y}^{\text{lin}}} =
$$

$$
\begin{bmatrix} a_{ss}u(k-N) \\ a_{ss}u(k-N+1) \\ a_{ss}u(k-N+2) \\ \vdots \\ a_{ss}u(k-N+M-1) \\ \vdots \\ a_{ss}u(k-N+P-1) \end{bmatrix} + \begin{bmatrix} a_2 & a_3 & \cdots & \cdots & a_N \\ a_3 & a_4 & \cdots & a_N & 0 \\ a_4 & a_5 & \cdots & 0 & 0 \\ \vdots & & & \vdots & \\ a_{M+1} & \cdots & \cdots & 0 & 0 \\ & \vdots & & & \\ a_{P+1} & \cdots & \cdots & 0 & 0 \end{bmatrix} \begin{bmatrix} \Delta u(k-1) \\ \Delta u(k-2) \\ \Delta u(k-3) \\ \\ \Delta u(k-N+1) \end{bmatrix}
$$

$$
\underbrace{}_{\mathbf{y}^{\text{past}}}
$$

$$
+ \underbrace{\begin{bmatrix} a_1 & & & 0 \\ a_2 & a_1 & & \\ a_3 & a_2 & a_1 & \\ \vdots & & & \\ a_M & a_{M+1} & \cdots & a_1 \\ \cdots & & & \\ a_P & a_{P-1} & \cdots & a_{P-M+1} \end{bmatrix}}_{\mathbf{A}} \underbrace{\begin{bmatrix} \Delta u(k) \\ \Delta u(k+1) \\ \Delta u(k+2) \\ \vdots \\ \Delta u(k+M-1) \end{bmatrix}}_{\Delta\mathbf{u}} + \underbrace{\begin{bmatrix} d(k+1) \\ d(k+2) \\ d(k+3) \\ \vdots \\ d(k+M) \\ \vdots \\ d(k+P) \end{bmatrix}}_{\mathbf{d}} \quad (10.40)
$$

or

$$
\mathbf{y}^{\text{lin}} = \mathbf{y}^{\text{past}} + \mathbf{A}\Delta\mathbf{u} + \mathbf{d} \tag{10.41}
$$

P denotes the length of the prediction horizon. M is the move horizon or the number of future moves $\Delta u(k), \ldots, \Delta u(k+M-1)$ calculated by the DMC algorithm discussed below. $\mathbf{A}$ is the dynamic matrix composed of the step response coefficients. The effects of the known past inputs on the future outputs is defined by the vector $\mathbf{y}^{\text{past}}$.

Often the future behavior of the disturbance is not known, and it is customary to assume that the future values will be equal to the currently estimated value, i.e., $d(k+i) = d(k)$ for $i = 1, \ldots, P$ and $d(k)$ is calculated from Eq. (10.39).

With these definitions, the future output can now be predicted from Eq. (10.41) for any given vector of future control moves $\Delta\mathbf{u}$.

Calculation of the Control Inputs The following optimization problem is used to calculate the control inputs:

$$\min_{\Delta u} \quad \sum_{i=1}^{P} \gamma^2(i)(y^{sp}(k+i)^{\cdot} - y^{lin}(k+i))^2$$

$$+ \quad \sum_{j=1}^{M} \lambda^2(j)(\Delta u(k+M-j))^2 \tag{10.42}$$

y^{sp} is the setpoint, and γ and λ are the time-varying weights on the output error and on the change in the input, respectively. The solution to the above problem is a least squares solution in the form of the following linear DMC control law:

$$\Delta u = (A^T \Gamma^T \Gamma A + \Lambda^T \Lambda)^{-1} A^T \Gamma^T \Gamma (y^{sp} - y^{past} - d) \tag{10.43}$$

where G and Λ are diagonal matrices containing the weights $\gamma(i)$ and $\lambda(i)$, respectively. Usually only the first calculated move $\Delta u(k)$ is implemented and the calculations are repeated at the next sampling time to account for changing disturbances and to incorporate feedback. P, M, Γ, and Λ are used as tuning parameters by the designer.

Nonlinear Model-Predictive Control (NLMPC) The nonlinear DMC algorithm proposed here stems from reinterpreting the disturbance vector, d, appearing in the prediction Eq. (10.40). If the linear DMC is implemented on a linear plant, the vector d includes the effects of external disturbances, assuming a perfect linear model. However, if the same control law is applied to a nonlinear plant (see Fig. 10.18), the vector d will contain contributions from nonlinearities defined as d^{nl} as well as external disturbances defined as d^{ext}, i.e.,

$$\begin{bmatrix} d(k+1) \\ \vdots \\ d(k+P) \end{bmatrix} = \begin{bmatrix} d^{ext}(k+1) \\ \vdots \\ d^{ext}(k+P) \end{bmatrix} + \begin{bmatrix} d^{nl}(k+1) \\ \vdots \\ d^{nl}(k+P) \end{bmatrix} \tag{10.44}$$

In linear DMC, the vector d and, therefore, the future disturbance vector due to nonlinearities, d^{nl}, is assumed constant. In the nonlinear DMC formulation, the vector d^{nl} will be estimated using a nonlinear dynamic model and will no longer be assumed constant. The development proceeds as follows.

First, in accordance with Eqs. (10.41) and (10.44), we define an *extended linear* DMC model over the future prediction horizon of length P:

$$y^{el} = y^{past} + A\Delta u + d^{ext} + d^{nl} \tag{10.45}$$

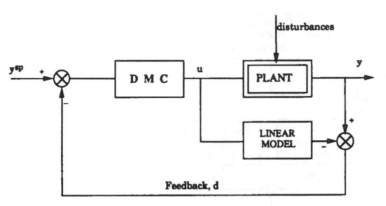

Figure 10.18   Linear DMC. (From Ref. 20.)

where $\mathbf{d}^{nl}$ is varying from one sampling time to another, and $\mathbf{d}^{ext}$ is assumed constant over the prediction horizon. The optimal DMC inputs for the extended linear model become

$$\Delta \mathbf{u} = (\mathbf{A}^T \Gamma^T \Gamma \mathbf{A} + \Lambda^T \Lambda)^{-1} \mathbf{A}^T \Gamma^T \Gamma (\mathbf{y}^{sp} - \mathbf{y}^{past} - \mathbf{d}^{ext} - \mathbf{d}^{nl}) \quad (10.46)$$

The objective is to compute a time-varying disturbance vector, $\mathbf{d}^{nl}$, that captures the future disturbance due to nonlinearities and renders the above $\Delta \mathbf{u}$ optimal for the *nonlinear* model. This is accomplished by requiring the output of the extended linear model $(\mathbf{y}^{el})$ match the output from the nonlinear model $(\mathbf{y}^{nl})$ at all the future sampling times:

$$\begin{bmatrix} y^{nl}(k+1) \\ \vdots \\ y^{nl}(k+P) \end{bmatrix} = \begin{bmatrix} y^{el}(k+1) \\ \vdots \\ y^{el}(k+P) \end{bmatrix} = \mathbf{y}^{past} + \mathbf{A}\Delta\mathbf{u} + \mathbf{d}^{ext} + \begin{bmatrix} d^{nl}(k+1) \\ \vdots \\ d^{nl}(k+P) \end{bmatrix}$$

$$(10.47)$$

This is a set of P simultaneous nonlinear equations in P unknowns, $\mathbf{d}^{nl}(k+1)$, ..., $\mathbf{d}^{nl}(k+P)$. The quantities $\mathbf{y}^{past}$ and $\mathbf{d}^{ext}$ can be separately calculated as before (see below), and $\Delta \mathbf{u}$ is obtained from Eq. (10.46). Note also that the terms containing nonlinearities with respect to $\mathbf{d}^{nl}$ appear in the $\mathbf{y}^{nl}$ vector. This is because the output from the nonlinear model will have nonlinear dependence on the input, which in turn depends on $\mathbf{d}^{nl}$ through Eq. (10.46).

Any standard nonlinear equation solution method can be used to obtain the vector $\mathbf{d}^{nl}$ satisfying Eq. (10.47). One way is by using a successive substitution or fixed point algorithm:

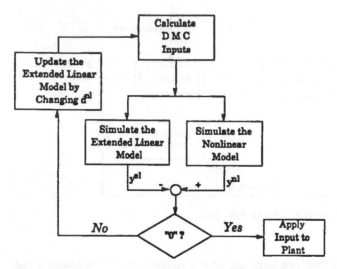

Figure 10.19 The flowchart of the nonlinear DMC algorithm. (From Ref. 20.)

$$d_{l+1}^{nl} = d_l^{nl} + \beta(y_l^{nl} - y_l^{el}) \tag{10.48}$$

where l is the iteration number; β is a relaxation factor in $(0,1)$ used to enlarge the region of convergence.

A flowchart for the nonlinear DMC algorithm showing the fixed point iterations is given in Fig. 10.19. Only the first control action $\Delta u(k)$ is implemented after the iterations [Eq. (10.48)] converge, and the iterative DMC calculations are repeated at the next sampling time. When it converges, the algorithm gives all the advantages of classical DMC. A control block diagram is shown in Fig. 10.20. The internal loop wrapped around the DMC block represents the fixed point iterations. At each sampling time, the switch on the control input closes with the linear model during iterations. Upon convergence, it opens and closes with the plant, and the cycle is repeated at the next sampling time.

Application to a Semibatch Polymerization Reactor

The NLMPC algorithm described above was applied, in simulation, to the control of a semibatch reactor for the polymerization of methyl methacrylate. This particular system was chosen because of its industrial importance, its strong nonlinearities, and because it illustrates the capabilities of NLMPC in handling control challenges unique to batch operations. In what follows, the reaction system will be briefly described, and the application of NLMPC to both single-loop and multivariable control of this reactor will be demonstrated.

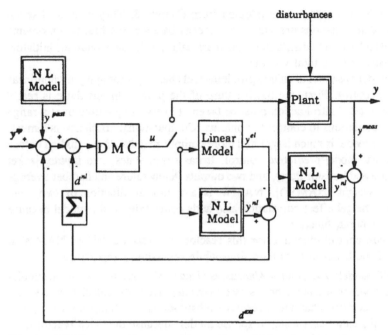

Figure 10.20 Block diagram of the nonlinear DMC algorithm. (From Ref. 20.)

Polymerization Reactor Model The system chosen for study is the semibatch, free radical polymerization of methyl methacrylate. The physical system consists of a jacketed reactor with inlet flow of initiator and solvent. The initiator used is benzoyl peroxide dissolved in ethyl acetate solvent. Methyl methacrylate monomer is charged to the reactor initially and is not added during the reaction. The manipulated variables are initiator flow, u_1, and cooling jacket temperature, u_2. The controlled outputs are the reaction temperature, y_1, and the number average molecular weight, y_2:

$$y_2 = \frac{\sum\limits_{i=2}^{\infty} M_i n_i}{\sum\limits_{i=2}^{\infty} n_i} \tag{10.49}$$

where M_i is the molecular weight of molecule i and n_i is the number of molecules with weight M_i.

The modeling equations are adopted from Chapter 3. They consist of seven nonlinear state equations describing the energy balance, the first two moments of the molecular weight distribution, the mass balances for the monomer, initiator, and solvent, and the total volume.

This model poses some unique problems and challenges for applying nonlinear predictive control. First, the batch nature of the process means there is really no steady state. Control action must be taken at every sample time over a range of reactor conditions to control the outputs. Computational efficiency is a must. Second, the model is much larger and more complicated than those studied before for nonlinear model-predictive control. It has seven states, two inputs (jacket temperature and initiator flow), and two outputs [temperature and number average molecular weight (NAMW)]. NAMW is a nonlinear function of two state variables. The gel effect correlation also adds complexity to the model because it is not a smooth function.

The main control objective for this reactor is to reach a desired NAMW at the end of the batch in minimum time while regulating temperature.

NLMPC Models and Control Algorithms Used in Simulations   In all the simulations to follow, the secant method is used in the iterative DMC calculations because it proved to be faster than the successive substitutions. For temperature control, the DMC algorithm uses a step response model. Because the batch reactor does not have a true "steady state," the step coefficients are determined as shown in Fig. 10.8. Keeping a constant input and integrating forward gives the unperturbed output y^{past}. Stepping the input to a new value and integrating into the future gives the perturbed output, y^{per}. The step response coefficients are then calculated from

$$a_i = \frac{y^{per}(k + i) - y^{past}(k + i)}{\Delta u} \tag{10.50}$$

In the case of molecular weight control, open-loop simulation revealed that the initiator had very different effects on the molecular weight depending on the time it was introduced to the batch reactor. This was best observed by adding an initiator *pulse* of 12 min duration (i.e., one sampling time) at different times during the reaction rather than introducing a step input which lasts over the entire time of reaction. Consequently, instead of the step model, an impulse model of the following form [26] is used in the DMC algorithm for the molecular weight:

$$y(k) = \sum_{i=1}^{N} h_i u(k - i) \tag{10.51}$$

The impulse coefficients, h_i, are calculated similar to the step response coefficients by integrating the nonlinear system with a pulse input instead of a step.

This similarity allows use of essentially the same secant method for the impulse response model.

For the temperature loop, we use $M = P = 1$. For the molecular weight loop, $M = 1$ and at each sampling time the prediction horizon is set equal to the final batch time, t_f, which is defined as the time it takes to reach 85% monomer conversion. Because t_f is not known a priori, it is calculated on-line and the prediction horizon (P) is updated accordingly at each sampling time. The NLMPC tries to match the molecular weight with its target setpoint only at the end of the batch. The batch terminates when 85% conversion is reached.

Finally, it is important to mention that the step (impulse) response coefficients were updated from one sampling time to another. After the secant iterations converge and the control $\Delta u(k)$ is implemented, the simulated nonlinear outputs are used to update either the step coefficient a_1 for the temperature control,

$$a_1 = \frac{y^{nl}(k+1) - y^{past}(k+1)}{\Delta u(k)} \tag{10.52}$$

or the impulse coefficient h_p for molecular weight targeting,

$$h_p = \frac{y^{nl}(k+P) - y^{past}(k+P)}{u(k)} \tag{10.53}$$

where $k + P$ denotes the end of the batch. Note that only one coefficient needs to be updated. This is because $M = 1$, and only one future value for each output is predicted ($y^{nl}(k+1)$ for for temperature and $y^{nl}(k+P)$ for molecular weight).

Single Loop Control   Often the control objective for a batch polymerization reactor is to make the temperature of the reactor follow some predetermined trajectory. In the following simulations, the temperature is regulated around its initial setpoint and ramped up by adjusting the jacket temperature.

Figure 10.21 shows the results of the NLMPC with the secant method applied with a sampling time of 12 sec, $M = P = 1$, and no plant/model mismatch for regulation around 340 K. Note that the temperature is perfectly controlled matching the setpoint at each sampling time. The intersample deviations appear as shaded areas due to small sampling time relative to the plotted time scale. Figures 10.22 and 10.23 show the control input (dimensionless jacket temperature) and conversion. Three distinct regions are visible.

In Region I, the reaction rate is decreasing slowly as reactants are consumed. The model predicts a reduced heat generation rate and increases the jacket temperature (see Fig. 10.22) which increases the reaction temperature during the first part of the sampling period. The decreasing reaction rate brings the temperature down to the setpoint at the beginning of the next sampling period. Because this is a batch reaction with constantly changing conditions and no real

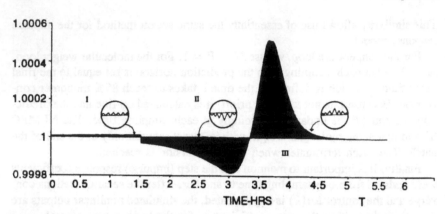

Figure 10.21 Dimensionless temperature using the secant algorithm. The actual temperature oscillations are at most a few tenths of a degree. (From Ref. 20.)

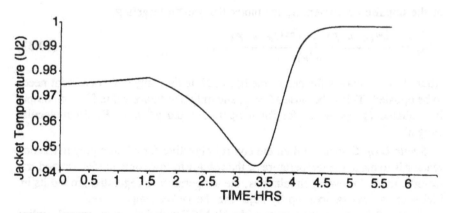

Figure 10.22 Dimensionless jacket temperature. (From Ref. 20.)

steady state, the intersampling oscillations of the output are unavoidable with a discrete controller.

In Region II, the gel effect causes the reaction to accelerate, creating more heat to be removed. Thus, a trend just opposite of Region I is observed. The intersample deviations are more pronounced in this region due to the gel effect. When the reactor is most exothermic (at 3 hr), if the calculated control action is kept constant for more than two sampling times, the reactor experiences a runaway. This makes frequent, precise control moves necessary to operate in this region.

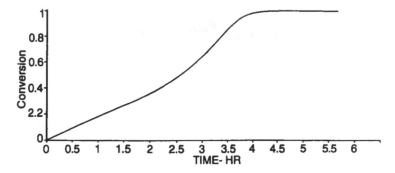

Figure 10.23 Monomer conversion. (From Ref. 20.)

In Region III, the depletion of reactants overcomes the gel effect and the reaction rate slows again. The intersampling oscillations follow the same trend as in Region I. When all of the monomer is reacted, the reactor and jacket are both at the same temperature. Generally, the reaction is stopped before reaching complete conversion. In this case, we have chosen to stop the reaction arbitrarily at 85% conversion. In all three regions, the actual magnitude of the intersample oscillations is insignificant. They are magnified by the choice of scale in Fig. 10.21 to illustrate the workings of the NLMPC.

The results for regulation along a temperature trajectory are shown in Fig. 10.24. The temperature is initially regulated at 340K. At 1 hr, the setpoint is ramped to reach 343K at 2 hr. After 2 hr, the temperature is regulated at 343K. The trajectory is followed perfectly, reaching the setpoint at the end of each 12-sec sampling time. Trends similar to Fig. 10.21 appear after 2 hr into the simulation. The temperature oscillates below the setpoint when the gel effect causes the reaction to become highly exothermic, and it oscillates above the setpoint when the reaction slows down due to monomer depletion.

Figure 10.25 shows the result of applying PID control to the batch polymerization reactor model with a sampling time of 12 sec and the same setpoint trajectory. Even though the controller is sluggish early in the polymerization, it eventually goes unstable after 2 hr when the reaction enters the extreme exothermic region. This is not surprising because calculations indicate that the system gain increases by approximately two and one-half orders of magnitude as the reaction enters the autocatalytic (gel effect) region.

Multivariable Control In this section, the NLMPC with the secant algorithm is used to solve control problems (for the semibatch polymerization reactor) associated with a combination of regulation and targeting of terminal (end of batch) conditions.

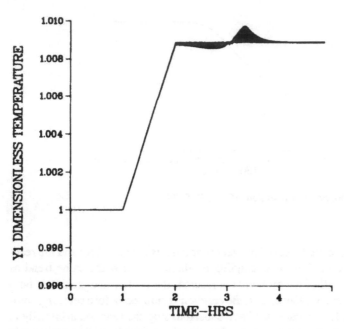

Figure 10.24 Temperature control with the secant algorithm. (From Ref. 20.)

Attempts to control average molecular weight with jacket temperature (with all initiators added at the beginning of the reaction) were unsuccessful because the reactor temperature is not directly controlled, and eventually goes unstable [26]. Therefore, one has to control molecular weight (y_2) *and* temperature (y_1) by manipulating the jacket temperature (u_2) *and* the initiator flow (u_1). Instead of developing a multivariable nonlinear DMC controller for the full 2×2 nonlinear input–output operator, a different approach which exploits the model structure is taken. Specifically, examination of the model reveals that the jacket temperature appears only in the state equation for the reactor temperature. Therefore, if the reactor temperature could be "perfectly" regulated, the molecular weight could be controlled by using the nonlinear *isothermal* mode. Temperature would then be controlled by using the full model with initiator calculated from the isothermal model. The block diagram is shown in Fig. 10.26.

The molecular weight and temperature controllers are two separate nonlinear DMC controllers. In the temperature loop, there is no plant/model mismatch because the model is the full perfect model. However, the molecular weight loop has a mismatch because control is based on the isothermal model, i.e., it is assumed that the temperature loop is perfectly regulating. As we have seen in the single-loop simulations, temperature is regulated tightly; therefore, this mismatch is not

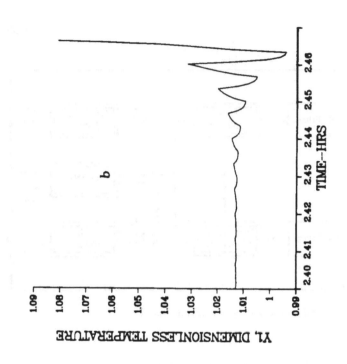

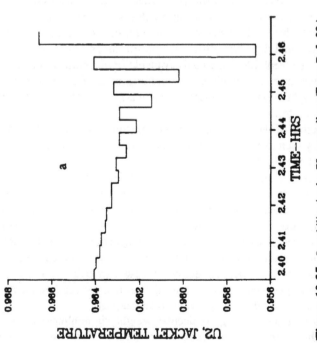

Figure 10.25 Instability in the PI controller. (From Ref. 20.)

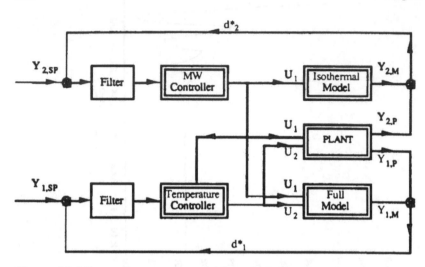

Figure 10.26 MIMO block diagram. (From Ref. 20.)

expected to be serious. In addition, first-order filters are incorporated into each loop to achieve robustness against any model/plant mismatches.

Finally, note that, with this control structure, different sampling times can be easily incorporated into the different control loops. For example, because molecular weight feedback is often only available from chromatographic techniques with long measurement deadtimes, a sampling time of 12 min is used for the molecular weight controller, whereas a sampling time of 9 sec is used for the temperature controller.

The objective of the molecular weight controller is to calculate the initiator feed trajectory such that the average molecular weight is equal to the desired value at the end of the batch. This can be accomplished by regulating the average molecular weight (i.e., keeping the average molecular weight at setpoint at all times during the batch) or by manipulating the inputs in such a way as to bring the average molecular weight to its target at the end of the reaction (targeting). The first alternative (regulation) will force the average molecular weight to be on target at all times, including the final time. Regulation may be desirable because it will provide a product which not only has the correct average molecular weight, but which also has a low polydispersity (narrow distribution). Targeting, on the other hand, will produce a product with the desired average molecular weight, but with a larger polydispersity (wider distribution).

When regulation of average molecular weight via nonlinear NLMPC is implemented, the controller first calculates the addition of the initiator to the reactor. Later, the controller calculates a negative initiator addition, i.e., removal

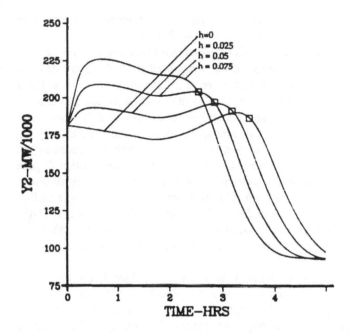

Figure 10.27 Response of MW to a 12-min pulse in initiator flow at time = 0. The NAMW at 85% conversion for each simulation is indicated by the square marker. The height of the pulse is h. (From Ref. 20.)

of initiator from the reactor, a physical impossibility. The unidirectional nature of the manipulation (initiator can be added, but not removed) makes regulation of average molecular weight by initiator feed impossible. Instead, the targeting problem is solved by the NLMPC using the secant method to predict and control the average molecular weight at the end of the batch. The end of the batch is defined for this study to be the time when the monomer conversion reaches 85%.

Figure 10.27 shows when each simulated batch reaches 85% conversion. It can be seen that the molecular weight at 85% conversion increases almost linearly with the height of the initiator pulse, indicating that the secant method should yield good results for this problem.

In the following results, the MIMO control structure in Fig. 10.26 is implemented. The secant algorithm is used in both controllers. A first-order filter is used in the molecular weight controller, whereas the temperature controller is unfiltered. In the molecular weight controller, the initiator pulse which causes the molecular weight to reach the filtered setpoint at the end of the batch (85% conversion) is calculated for a 12-min sampling time. In other words $M = 1$ and

P is set equal to the time it takes for 85% conversion. The temperature controller predicts and corrects one step ahead with $M = P = 1$.

Figure 10.28 shows the results of this MIMO control structure with a molecular weight setpoint of 195,000 and the temperature regulated at 340K. Figure 10.28(a) shows the dimensionless reactor temperature as a function of time. In addition to the previously described oscillations due to monomer depletion and autoacceleration in single-loop temperature control (refer to Fig. 10.21), a new relationship with initiator addition can also be seen. As an initiator is added by the molecular weight controller, the reaction rate increases, making it necessary to decrease the jacket temperature to keep the reaction temperature constant. Figure 10.29(a) shows that the slope of the jacket temperature decrease is approximately proportional to the initiator addition. This also increases the intersample oscillations in size [compare Region I in Figs. 10.21 and 10.28(b)]. Figure 10.29(b) shows that the desired molecular weight at 85% conversion (195,000) was achieved. Similar results are obtained at other temperature and average molecular weight setpoint.

These simulations show that NLMPC with the secant algorithm is very effective in calculating inputs for temperature and molecular weight control. In the absence of model/plant mismatch though, this control algorithm is nothing more than the calculation of an open-loop input profile. The utility of the proposed method can be clearly illustrated when the process is simulated with an inhibitor disturbance to show the feedback properties of the algorithm in the presence of uncertainty.

An inhibitor is added to the monomer to prevent thermal initiation and polymerization during transit and storage. This inhibitor may or may not be removed before polymerization. Either way, the level of inhibition for a given lot of monomer is often unknown. An imprecise estimate of the inhibitor level can distort the expected open-loop molecular weight response. Figure 10.30 shows the results of the secant algorithm with and without molecular weight feedback. To simulate the effect of residual inhibitor on the polymerization, the initiator efficiency of the *plant* was assumed to be 0.1 for the first two sampling intervals (24 min). After two sampling periods, the inhibitor was assumed to have been consumed in reaction with the initiator, and the initiator efficiency was set to 0.5 for the rest of the polymerization. The initiator efficiency of the *model* remained constant at 0.5. Figure 10.30 shows three molecular weight profiles. Profile 1 shows control at a temperature of 343K and an average molecular weight setpoint of 175,000 with *no inhibitor disturbance*. Profile 2 shows what happens when there is an inhibitor disturbance and the control inputs are not updated. A different profile results with the final molecular weight about 2000 units over setpoint. Profile 3 shows the effect of disturbance feedback. With corrective feedback, the molecular weight profile is brought back to follow a path similar to profile 1. The average molecular weight at 85% monomer conversion is at the

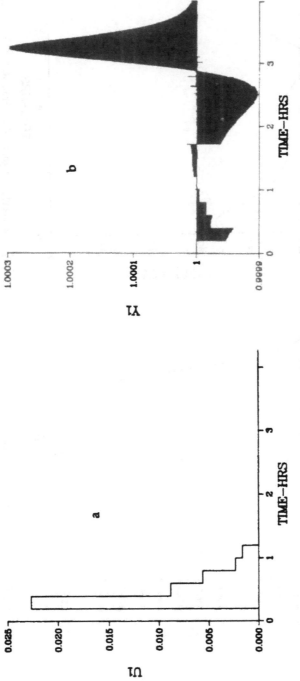

Figure 10.28 MIMO simulations. (a) Shows the initiator addition; (b) shows the dimensionless temperature. (From Ref. 20.)

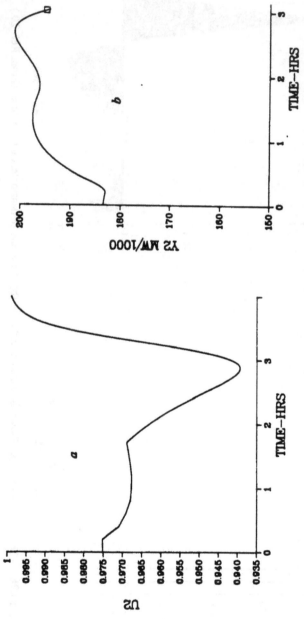

Figure 10.29   MIMO simulations. (a) Shows the jacket temperature; (b) shows the molecular weight profile. (From Ref. 20.)

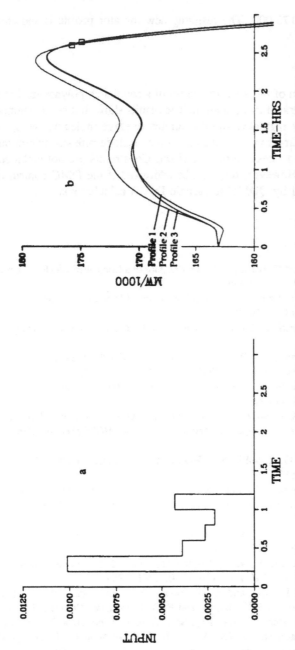

Figure 10.30 The result of molecular weight feedback. Profile 1 is for control with a perfect model. Profile 2 indicates the results when the outputs from profile 1 are applied to a model with an inhibitor disturbance. Profile 3 shows feedback correction of the disturbance. (a) Shows the initiator associated with Profile 1. (From Ref. 20.)

desired value of 175,000. The resulting new initiator profile is shown in Fig. 10.30(a).

Conclusions

The application of NLMPC to control of a semibatch polymerization reactor has been demonstrated in simulation. The results show that reactor temperature can be effectively regulated and the number average molecular weight can be forced to meet a target value at the end of the batch despite strong nonlinearities due to the reaction rates and the gel effect. Constraints are not included in the present method. However, they can be addressed if the DMC control calculations are replaced by QDMC (quadratic DMC) calculations.

References

1. L. Pellegrini and G. Biardi, "Chaotic Behavior of a Controlled CSTR," *Comp. Chem. Eng.*, *14*, 1237–1247 (1990).
2. B. W. Bequette, "Nonlinear Control of Chemical Processes, A Review," *Ind. Eng. Chem. Res.*, *30*, 10 (1991).
3. A. Isidori, *Nonlinear Control Systems: An Introduction*, 2nd ed., Springer-Verlag, Berlin, 1989.
4. C. Kravaris and C. B. Chung, "Nonlinear State Feedback Synthesis by Global Input/Output Linerization," *AIChE J.*, *33*, 592–603 (1987).
5. A. J. Boye and W. L. Brogan, "A Nonlinear System Controller," *Int. J. Contr.*, *44*, 1209–1218 (1986).
6. F. B. Bartee, K. F. Bloss, and C. Georgakis, "Design of Nonlinear Reference System Control Structures," paper presented at the Annual *AIChE* Meeting, San Francisco, CA 1989.
7. P. L. Lee and G. R. Sullivan, "Generic Model Control," *Comp. Chem. Eng.*, *12*, 573–580 (1988).
8. C. G. Economou, M. Morari, and B. O. Palsson, "Internal Model Control. 5. Extension to Nonlinear Systems, *Ind. Eng. Chem. Proc. Des. Dev.*, *25*, 1986, pp. 403–411.
9. B. D. Kulkarni, S. S. Tambe, N. V. Shukla, and P. B. Deshpande, *Nonlinear pH Control*, *46*, 995–1003 (1991).
10. R. A. Wright and C. Kravaris, "Nonlinear Control of pH Processes Using the Strong Acid Equivalent," *Ind. Eng. Chem.* 30, 1561, 1991.
11. A. Uppal, W. H. Ray, and A. B. Poore, *Chem. Eng. Sci.*, *29*, 967 (1974).
12. V. Ravi Kumar, B. D. Kulkarni, and P. B. Deshpande, "On the Robust Control of Nonlinear Systems," *Proc. Roy. Soc., London, Series A*, *433*, 711–722 (1991).
13. J. K. Bandyopadhayay, V. Ravi Kumar, B. D. Kulkarni, and P. B. Deshpande, "On Dynamic Control of Chaos," *Phys. Lett. A.*, 166, 197 (1992).
14. C. Kahlert, O. E. Rossler, and A. Verma, Springer Series in Chem. Phys. No. 18, 1981, p. 355.

15. A. K. Adebekun and F. J. Schork, "Continuous Solution Polymerization Reactor Control. 1. Nonlinear Reference Control of Methylmethacrylate Polymerization," *I & EC Research*, *28*, 1308–1324 (1889).
16. A. J. Boye and W. L. Borgan, "A Non-linear System Controller," *Int. J. Control*, *44*, 1209–1218 (1986).
17. R. D. Bartusiak, C. Georgakis, and M. J. Reilly, "Designing Nonlinear Control Structures by Reference System Synthesis," *Proc. Amer. Cont. Conf.*, 1585–1590 (1988).
18. B. Porter and H. M. Power, "Controllability of Multivariable Systems Incorporating Integral Feedback," *Electron Lett.*, *6*, 6–7 (1970).
19. A. K. Adebekun and F. J. Schork, "Continuous Solution Polymerization Reactor Control. 2. Estimation and Nonlinear Reference Control During Methylmethacrylate Polymerization," *I & EC Research*, *28*, 1846–1861 (1989).
20. T. Peterson, E. Hernandez, Y. Arkun, and F. J. Schork, "A Nonlinear DMC Algorithm and Its Application to a Semi-Batch Polymerization Reactor," *Chem. Eng. Sci.*, *42*, 737 (1992).
21. T. J. McAvoy, Y. Arkun, and E. Zafiriou (eds.), *IFAC Workshop Proceedings on Model Based Process Control*, Pergamon Press, Elmsford, NY, 1989.
22. D. M. Prett and C. E. Garcia, 1988, Fundamental Process Control, Butterworths, 1988.
23. D. D. Brengel and W. D. Seider, "A Multi-Step Nonlinear Prediction Controller," ACC Proc. 1100–1105 (1989).
24. J. W. Eaton, J. B. Rawlings, and T. F. Edgar, "Model Predictive Contol and Sensitivity Analysis for Constrained Nonlinear Processes," in *IFAC Workshop Procedings on Model Based Control* (T. J. McAvoy, Y. Arkun, and E. Zafiriou (eds.), Pergamon Press, Elmsford, NY, 1988.
25. W. C. Li and L. T. Biegler, "A Multi-Step, Newton Type Strategy for Constrained Nonlinear Processes, *ACC Proc.* 1526–1527 (1989).
26. T. J. Peterson, "Nonlinear Predictive Control by an Extended DMC," M.S. thesis, Georgia Institute of Technology, Atlanta, GA, 1990.

11

Introduction to Polymer Processing

Vikas M. Nadkarni

National Chemical Laboratory
Pune, India

Polymers are firmly established as a prime class of materials of significant economic relevance because of the possibility of converting the macromolecules into useful products and functional articles by a variety of processing techniques. In this chapter, the different methods used for processing thermosetting polymers, thermoplastics, and polymeric composites are reviewed. The influence of molecular structure on processability of polymers is highlighted to underscore the importance of process control in polymerization to ensure consistent processing behavior.

The effects of processing variables on the microstructure of the finished product in the solid state are briefly discussed. This processing structure property interdependence in polymers would then form the basis for outlining the framework of the control problem in polymer processing.

11.1 Technological Perspective

Chemists invented the world of polymers when they started combining small organic molecules to form macromolecules. Depending on how they were put together chemically, macromolecules became resins, plastics, or elastomers. When engineers started converting these macromolecules to useful products, they realized that they could make these materials work as textile fibers, packaging films, tapes, records, lighting fixtures, foam cushions, toys, electrical connectors, gaskets, bearings, auto tires, luggage, airplane parts, artificial limbs, and so on. Thus, polymers became synonymous with adhesives, coatings, synthetic fibers, foams, and composites. As a result, polymers have been steadily replacing conventional materials such as jute, paper, wood, glass, cement, and metals in diverse applications over the past 50 years (Tables 11.1 and 11.2). It is this broad spectrum of seemingly diverse forms and applications in a variety of markets that has led to the emergence of the "Polymer Industry."

The world consumption of polymeric materials is of the order of 50 million tons. The phenomenal growth of polymeric materials is attributable not only to their superior performance in terms of chemical resistance and light weight but also to the possibility of "tailor-making" these materials to suit specific end-use requirements. Historically, the major developments of industrial significance have taken place only since the 1930s, although cellulose nitrate and phenolic resins were introduced at the beginning of the nineteenth century. The initial developments in the 1930s and 1940s mainly concerned thermosets (unsaturated polyester, epoxies), bulk commodity thermoplastics (polyvinyl chloride, polystyrene, low-density polyethylene), and, of course, the invention of synthetic fibers from nylon. The 1950s and 1960s saw the advent of high-performance engineering plastics such as polyacetals, polycarbonate, polyphenylene oxide (Noryl), thermoplastic polyesters, and specialty polymers such as polysulfone, polyphenylene sulfide, fluoropolymers, etc. In all these developments, materials technology *preceded* processing technology and innovation in applications. Whereas in the last two decades, the phenomenal growth of the polymer industry has taken place because of the *concurrent* developments in materials technology, processing techniques, machinery, applications research, and product design. The developmental approach has also changed from empirical experimentation to "tailor-making" or engineering of materials to specific end-use requirements through multicomponent polymer systems such as fiber-reinforced composites, mineral-filled compounds, alloys, multilayer films and laminates, foams, etc. Thus, the 1970s and 1980s have given us novel materials like polymer blends and alloys, interpenetrating networks, thermoplastic elastomers, and liquid crystal polymers; microprocessor controlled machines; novel processing techniques like structural foam molding, multilayer extrusion and blow molding, reinforced reaction injection molding; compounding technology; computer-aided materials selection, product design, and mold design.

All these developments are not only concurrent and interdependent, but they also have had a synergistic effect on widening the horizons of the Polymer Industry. Over the past two decades, polymeric composites have emerged as engineering materials in their own right, for structural and also dynamic load-bearing applications. They have replaced metals in engineering applications such as gears, pump impellers, bearings, valve seats, seals, auto bumpers and fenders, springs, etc. High-performance composites employed in automotive, aerospace,

Table 11.1 Replacement of conventional materials by plastics.

Application	Polymers used	Conventional materials replaced
• Packaging		
Milk	LDPE film pouches	Coated paper, metal
Beverages	PET	Glass
Beverage crates	HDPE, PP	Wood
Pharmaceuticals	PVC, HDPE, PP	Glass, paper
Squeeze tubes (toothpaste, ointment)	HDPE, LLDPE, PP	Metal
Consumer goods like soaps	LDPE, PVC Shrink wrap	Paper
Retail carrier bag	LLDPE, HDPE	Paper
Pallet wrap	LDPE, LLDPE	Heavy duty paper
Fertilizer	LDPE heavy duty sacks	Jute
Cement	HDPE and PP woven sacks	Jute
Minerals	LDPE	Jute
Strappings	HDPE, PP, PET	Coir, jute, steel
Ropes	Nylon, PP, PET	Coir, sisal, jute, steel
Cushioning material	EPS, PU, PP, PE foams	Flannel cloth, paper
• Storage tanks		
Water	HDPE, LLDPE (rotomolded)	Cement, mild steel
Septic tanks	HDPE, LLDPE	Cement
Agrochemicals	HDPE	Steel
Automotive fuel	HDPE, PP	Steel
• Textile apparels	PET, nylon, acrylic	Cotton, rayon, silk
• Upholstery fabrics	PP, PVC, rexene	Cotton, rayon, jute
• Carpets	Nylon, acrylic, PP	Coir, jute
• Furniture	Phenolic laminates, HIPS, PP structural foam, FRP, PP canes	Wood, cane, bamboo
• Toys	PP, ABS, HIPS	Wood, sheet metal
• Luggage	PP (impact modified), ABS, FRP	Wood, steel
• Pipes, tubings	PVC, HDPE, LDPE, PP, FRP	Cement, steel
• Window panes, cabinet	Acrylic, polycarbonate	Glass
• Shoe soles	PVC, PP, structural foam, PU, Hypalon	Leather, rubber
• Buckets, tumblers	HDPE, LLDPE	Sheet metal, brass
• Window and door frames	PVC, FRP	Wood, steel, aluminum
• Sign boards	FRP, acrylic	Wood, metal
• Structural sections	Pultruded FRP	Steel, aluminum
• Appliance housings	ABS, FRP, PP, structural foam	Metal
• Plumbing fixtures	Polyacetals, ABS, PP, HDPE	Stainless steel, brass
• Textile cones and bobbins	PP, FRP	Cardboard

Table 11.2 High-performance engineering applications.

Gears	PPS, TPE, Polyacetals
Pump impellers	PET, PBT, nylons
Bush and ball bearings	PTFE, nylon/MoS$_2$
Springs for forming presses	PU elastomers
Automotive bumpers	PU, PP, alloys
Automotive suspension springs	Fiberglass/epoxies
High-speed train driver cabin	FRP/PVC foam
Turbine blades	Carbon fibers/epoxies
Ocean mining ropes	KEVLAR/PP composite
Fire fighter suits	Nomex fibers
Magneto in scooter engine	Polymeric magnet
Membrane switches (piezoelectric)	PVDF film
Automotive Engine	Polyimide

and defense applications use polymers as the matrix material. For example, fiberglass-reinforced epoxy suspension springs have been successfully tested for cars; metal pipes are being replaced by filament-wound FRP pipes in oil exploration; bulletproof vests can now be prepared from man-made lightweight aramid fibers; composite ropes made of aramid and polypropylene filaments are used for mining ocean floors where steel wires fail. Besides the consumer and engineering applications, novel separation technologies for dialysis, desalination, gas separation, etc., have become possible because of the development of suitable polymeric membranes; the developments in electronics have come about because of the availability of polymers for lithography and thin-film coating process technology; polymers are also being used for controlled release agrochemicals and medicines; the concept of barrier packaging through process innovation has extended the shelf life of foods considerably; in the past 50 years, synthetic fibers have steadily replaced cotton as the main material for clothing.

11.2   Polymeric Materials

The versatility of polymeric materials arises from the fact that one can "engineer" the structure through interaction of the processing and materials technologies. There are two broad classes of polymers, namely, thermosets and thermoplastics.

Thermosetting polymers are in viscous liquid or highly viscous paste form under ambient conditions before processing. During processing, they are transformed into hard, infusible solids through the formation of three-dimensional networks via chemical reactions under the influence of heat and pressure. The reactions are catalyzed chemically, thermally, or by radiation. Examples of thermosetting polymers are phenol-formaldehyde, urea-formaldehyde, melamine-formaldehyde, epoxies, and unsaturated polyester resins, polyurethanes, etc. High-performance thermosets such as isophthalate polyesters, epoxies, polyimides, polyurethanes, and polyureas are used as matrix materials for composites reinforced with fiberglass, carbon fibers, or aramid fibers.

Thermoplastics represent the predominant class of polymers, constituting 85–90% of the consumption of polymers, because of their use in large-volume applications such as packaging films, synthetic fibers, molded bottles and containers, and other articles. Thermoplastics are fusible solids under ambient conditions and are melted into viscous liquids for processing, followed by resolidification into the desired shape or form. The thermoplastics could be further categorized into four groups on the basis of their volume of production, cost, and thermal performance level, as indicated below:

1. High-volume/low-price commodity plastics. These include addition polymers, such as low-density polyethylene, high-density polyethylene, linear low-density polyethylene, polypropylene, polystyrene, polyvinyl chloride, etc., representing almost 75% of the thermoplastic consumption. These plastics can be used up to a maximum temperature of 100 °C.
2. Medium-volume/medium-price transition plastics. Representative examples of this group include ABS, polyacetals, polymethyl methacrylate, styrene copolymers, vinyl copolymers, etc., and they represent a transition group from commodity to engineering and specialty polymers.
3. Low-volume/high-price engineering plastics. This category consists of condensation polymers, such as nylon and polybutylene terephthalate (PBT), polyethylene terephthalate (PET), nylon 66, polycarbonate, polyphenylene oxide, etc. All of these polymers have been specially developed for metal replacement in engineering applications and offer thermal performance level between 150 and 200 °C. Nylons and PET are extensively used as synthetic fibers.
4. High-price specialty polymers. The examples of this category include polyphenylene sulfide, polyaryl ether ketones, polyimides, polysulfones, polyether-imides, polyarylates, and fluoropolymers such as PTFE, PVDF, ECTFE, etc. These specialty polymers can be used at elevated temperatures from 200 °C.

The performance level and cost-effectiveness of both the thermosetting and thermoplastic polymers can be improved by compounding them with reinforcing fibers such as fiberglass, carbon and aramid (Kevlar®) fibers, and/or with mineral fillers, such as mica, talc, wollastonite, etc. Property improvement is also possible via blending and alloying of thermoplastics and by the technique of interpenetrating networks (IPNs) of thermosets. A number of polymers can be converted in foamed

Table 11.3 Polymeric materials.

THERMOPLASTICS
1. Commodity Thermoplastics:
   Low-density polyethylene (LDPE)
   High-density polyethylene   (HDPE)
   Linear low-density polyethylene (LLDE)
   Polypropylene (PP)
   Ethylene-propylene copolymer
   Polyvinyl chloride (PVC)
   Polystyrene (PS)
   High-impact polystyrene (HIPS)

2. Transition Thermoplastics:
   Styrene-acrylonitrile copolymer (SAN)
   Polymethyl methacrylate (PMMA)
   Acrylonitrile-butadiene–styrene (ABS)
   Polyacetals (POM)

3. Engineering Thermoplastics:
   Nylon 6
   Nylon 66
   Poly(butylene terephthalate) (PBT)
   Poly(ethylene terephthalate) (PET)
   Polycarbonate (PC)
   Poly(phenyl oxide) (PPO)
   Polysulfones
   Poly(ether sulfones) (PES)

4. Speciality Thermoplastics:
   Polyphenylene sulphide (PPS)
   Poly(tetra fluoro ethylene) (PTFE)
   Poly(ether ether ketone) (PEEK)
   Poly(ether imide) (PEI)
   Polyimides
   Fluoroelastomers

5. Thermoplastic Elastomers:
   Styrene-isoprene-styrene (SIS)
   Ethylene-propylene-diene monomer (EPDM)
   Hard/soft block polyester ("Hytrel")
   Thermoplastic polyurethane (TPU)

THERMOSETS
   Phenolics (phenol-formaldehyde)
   Epoxies
   Urea-formaldehyde
   Melamine-formaldehyde

Table 11.3 Continued

Unsaturated polyesters
Polyurethanes
Silicones
Polyimides

ELASTOMERS
  Natural rubber
  Isoprene
  Neoprene
  Hypalon
  Butyl rubber
  Nitrile rubber

products by inert gas or chemical blowing during processing. Soft foam sheets of PE and PP are thus used for packaging electronic components, whereas molded rigid shapes of polystyrene foam are used for shock-proof packaging of instruments. PVC foam is used as a core for rigid composite laminates. Polyurethane flexible foams are used as seat cushions and PU rigid foams for insulation.

The different thermoplastic and thermosetting polymers are listed in Table 11.3. The variety of material forms of these polymers that can be obtained through processing are summarized in Table 11.4.

11.3 Polymer Processing Techniques

The stages in processing of thermosetting and thermoplastic polymers are quite different as illustrated schematically in Fig. 11.1. The processing of thermosets involves a chemical reaction, whereas thermoplastics processing involves only phase transformations. The various techniques of processing polymers may be classified into four categories:

• Melt processing	Thermoplastics
• Solid-state processing	Thermoplastics
• Solution processing	Thermoplastics
• Reactive processing	Thermosets

11.3.1 Melt Processing of Thermoplastics

The most common and economic technique of processing thermoplastics involves melting of the solid polymer pellets into a plasticating extruder followed by delivery

Table 11.4 Polymeric material forms.

Synthetic fibers
Films/sheets
Multilayer films/barrier packaging
Laminates
Pipes/tubes/profiles
Bottles/hollow shapes
Molded components
Foams
Separation membranes/"molecular filters"
Reinforcing fiber precursors
Composites
Polymeric magnets
Artificial organs/"biopolymers"

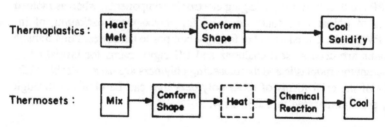

Figure 11.1 Stages in processing of thermosetting and thermoplastic polymers.

of the mechanically and thermally homogeneous melt to a tool for the shaping operation, with subsequent solidification to freeze the shape or the form. The tool is a die for extrusion or a mold for the molding methods. The main methods of melt processing are:

- Blown film extrusion
- Sheet extrusion
- Film casting/calendering
- Profile extrusion (pipes, tubes, c-sections, etc.)
- Injection molding
- Blow molding
- Melt spinning of synthetic fibers

Extrusion Processes

The extrusion processing methods are used for producing films of polyethylenes, polypropylene, and PVC for packaging; vinyl sheets for upholstery

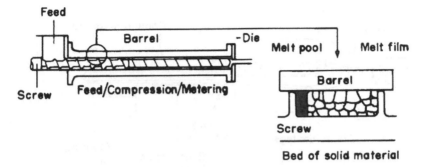

Figure 11.2 The three sections of an extruder screw, with a detail showing the melting mechanism in the feed section.

and wall covering; PET films for audio and video tapes; PVC and HDPE pipes for irrigation, domestic plumbing, sewage, natural gas transport, etc; PVC profiles for sidings, window frames; polycarbonate profiles for skylight roofing, etc.

The first operational stage in extrusion processing is the formation of the thermally and mechanically homogeneous melt. The dry polymer pellets are fed from a hopper into the extruder. The length of the extruder screw sections is divided into three sections termed feed, compression, and metering sections as schematically illustrated in Fig. 11.2 [1]. In the feed section, the pellets are plasticized to form a melt pool within the "screw flight." The molten material is carried forward by the rotating screw. Because the depth of the screw flight, defined as the difference between the screw shaft and the barrel diameters, is progressively reduced along the screw length, the melt is compressed in the second section. As a result of this compression and consequent viscous heating, a uniform melt temperature is ensured. The function of the final metering section of the screw is to deliver the melt at a constant rate to the die.

The extruder screw design is governed by the thermal and melt flow behavior of the polymer. The key design parameters for specifying the extruder machine are the length-to-diameter ratio of the barrel and the compression ratio of the screw, defined as the ratio of the maximum and minimum flight depths. Extruders with L/D ratios of 25–30 and high compression ratio are used for processing highly viscoelastic melts of addition polymers such as polyethylenes and polypropylene, whereas low L/D and low compression screw extruders are suitable for processing relatively less viscous melts of condensation polymers, such as nylons and thermoplastic polyesters.

Besides the screw design, the design of the extrusion die also depends on the melt rheology of the polymer. Different die geometries are employed for the different extrusion processes. Thus, blown film extrusion requires an annular die

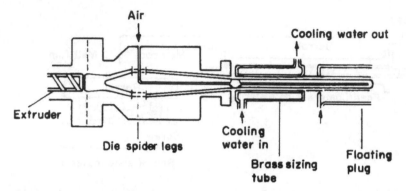

Figure 11.3 Die and cooling sections in an extrusion line for the manufacture of plastic pipe.

with vertical extrusion, whereas the sheet and cast film extrusion dies are called coat hanger dies. In pipe extrusion, the die is designed with a "torpedo" to split the melt delivered by the extruder the into annular shape; the torpedo is supported by the spider legs, as illustrated in Fig. 11.3 [1]. The main objective of the die design in any extrusion operation is to ensure delivery of the melt at a constant rate and uniform temperature at all points along the flow cross section. A variation in the temperature or delivery rate of the melt across the die length or periphery results in product defects such as fish eyes and streaks in the film.

The setting of a proper temperature profile along the screw barrel length is also critical for trouble-free operation and consistent product quality. Generally, there are three to five heating zones in the barrel, each with a separate heater and temperature controller. The zonal temperature profile is determined with reference to the thermal characteristics of the polymer. The maximum temperature of the melt, accounting for temperature rise due to viscous heating, has to be maintained below the thermal degradation temperature of the polymer.

Once the melt is extruded from the die, the shaping operation involves controlled solidification. The post-extrusion shaping operation could be as simple as casting of the molten extrudate sheet onto a chilled roll or, vacuum or internal air pressure sizing with quenching, as employed in pipe extrusion (Fig. 11.3). In profile extrusion, the final product shape and dimensions are achieved in the solidification state, in here as in cast film extrusion. The solidified film may be subjected to further uniaxial or biaxial stretching. In blown film extrusion, the stretching and solidification is achieved in a concurrent operation by blowing air from inside of the annular extrudate forming a bubble. The stretch ratio is governed by the bubble diameter, controlled mainly through the quench air temperature and flow rate. The properties of the extruded films are greatly influenced by the process variables employed in both the extrusion and shaping operations.

Molding Processes

One of the major reasons for the phenomenal growth of plastics in engineering applications has been the design flexibility offered by the molding processes to produce parts of complex shapes and geometries at high rates of production. The most common molding processes are injection molding, blow molding, and rotational molding. The injection molding process is used for large-volume production of components such as gears, lighting fixtures, valves, impellers, appliance housings, small toys, plumbing fixtures, etc., mainly with transition and engineering plastics. Blow molding produces hollow plastic containers ranging in capacities from baby milk bottles of a few ounces capacity to large 55-gal drums used for chemicals. The most commonly used materials for blow molding include commodity plastics such as high density polyethylene, PP copolymers, PVC, and engineering plastics such as polyethylene terephthalate and polycarbonate. The rotational molding technique is used for producing hollow articles of complex shapes and large volume, such as water storage tanks, rocking horse toys, automotive fuel tanks, etc., mainly with HDPE and LLDPE.

The injection molding operation starts with the plasticating extruder as the first stage, as in extrusion processing. However, unlike extrusion wherein the melt is delivered to the die in a continuous manner, the melt delivery or injection into the mold is an intermittent operation in injection molding. Thus, the extruder screw has to be designed for both rotational and reciprocating movements so that it can also act as a piston for melt injection. The sequence of operations involves:

- Melting, plasticization and metering in the extruder
- Melt accumulation in front of the screw
- Melt injection through the nozzle into the mold cavity
- Mold filling through sprue, runners and gates
- Mold closing and cooling for solidification
- Demolding and ejection of the part

The cycle of operations and equipment schematic are shown in Fig. 11.4 [1]. Some of the operations are carried out simultaneously to minimize cycle time. For example, the solidification of one shot of molding occurs at the same time a new batch of melt is being prepared in the extruder.

The design of the molds is made with reference to the flow and crystallization characteristics of the polymer to ensure complete mold filling before solidification and a stress-free, dimensionally stable product after demolding. In multiple cavity molds, the melt flow length from the nozzle to every cavity has to be equal so that a fixed shot weight is delivered to every cavity. In the case of a number of semicrystalline thermoplastics, the mold temperature needs to be controlled

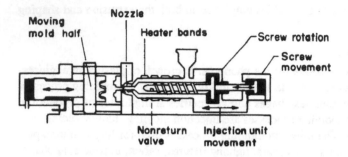

Key : ■ hydraulic system   ▭ polymer

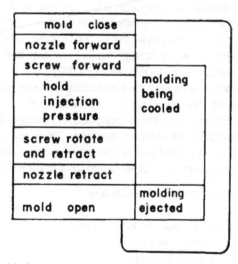

Figure 11.4 An injection molding machine with hydraulic mold closing, showing the cycle of operations.

to obtain consistent quality, warpage-free parts. Thus, cooling channels need to be incorporated in the mold block for water or oil circulation.

The critical process variables influencing the structure in the solid state and hence the properties of the molded product are the melt and mold temperatures, injection velocity and pressure, holding pressure, and cooling time. The "process time" above the T_g of the polymer in the mold cavity has to be longer than the stress relaxation time for amorphous polymers and the crystallization time

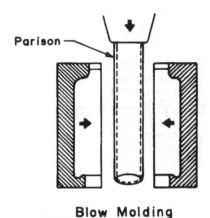

Blow Molding

Figure 11.5 Blow-molding process.

for semicrystalline polymers, to minimize residual stresses and to ensure complete crystallinity development.

The two basic blow molding processes are extrusion blow-molding and injection blow-molding. Both these processes involve three operations: forming of a hollow tube, known as a parison; placing the parison between halves of a forming mold; and blowing the parison with inert gas pressure from inside so that it assumes the shape of the mold. It is the process of forming the parison that is different in the two blow-molding processes. In injection blow-molding, the parison with the neck finish completely formed is molded in an injection mold, whereas in extrusion blow-molding the parison is extruded as a hollow tube. The process is schematically illustrated in Fig. 11.5.

It may be noted that the parison blowing operation involves simultaneous stretching and solidification as in blown film extrusion. Thus, the molten polymer is subjected to both shearing and extensional flow fields in blow-molding, whereas injection molding involves shearing flow of the molten polymer. The implications of these differences in flow fields on structure development are discussed in a subsequent section.

The rotational molding process uses polymer powder as the starting material rather than the pellets and it is used for producing large-volume hollow containers and hollow articles of complex shapes. The process involves loading of a weighed quantity of the polymer powder into the mold. The mold is then transferred to an oven. In the oven, the charge is heated and simultaneously revolved slowly around two or more axes, causing the polymer to distribute evenly over the inside surface and coat the mold. The mold is gradually heated to temperatures above the melting point of the polymer, thereby melting the well-distributed

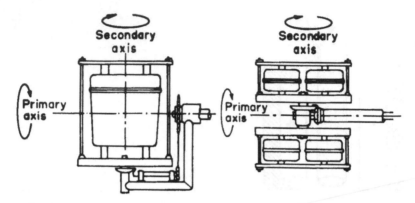

Figure 11.6 Two-axis rotational molding machine. Offset arm model is shown at left, straight arm at right.

polymer powder along the mold walls. Hot air is the common heating medium. After providing sufficient residence time for complete fusion, the mold is moved to a cooling area and air-cooled by large fans while being rotated. The part is then demolded. Figure 11.6 [2] schematically illustrates a two-axis rotational molding machine. The polymers most commonly processed by rotational molding are HDPE and LLDPE. In view of the severity of the processing conditions such as high temperatures, long cycle time, and oxidative atmosphere, the rotomolding grades of the various polymers need to be compounded with additives including heat stabilizers, antioxidants, etc. Moreover, proper optimization and control of the processing conditions is required to ensure good and consistent mechanical performance of the molded part particular for avoiding stress cracking.

Melt Spinning of Synthetic Fibers and Tapes

Synthetic fibers, filaments, and tapes made from polyethylene terephthalate, nylons, polypropylene, and high-density polyethylene are produced by the process of melt spinning followed by drawing and annealing. Clothing made of PET texturized filament yarns and staple fiber yarns of PET/cotton blends have replaced natural fibers in apparel and linen applications; nylon fabrics are used in umbrellas, parachutes, tents, and flexible luggage: nylon tire chords are used for rubber reinforcement; nylon monofilament are used as bristles on brushes; PP and HDPE filaments are used for making twine and high-strength marine ropes; PET and PP strappings have replaced steel strappings for packaging; texturized PP filaments are used for upholstery fabrics. There are a number of other consumer and industrial applications of synthetic fibers and tapes.

The first stage in the melt spinning process is again the melting of the polymer pellets in an extruder for delivering the melt to a shaping tool. The shaping tool in this case is termed a spinneret and it consists of a circular or rectangular plate with a number of fine holes through which the melt is extruded. The number of holes per spinneret are of the order of 35–100 for filament products and could be as high as 2000 for staple (short) fiber production. Because the capillary diameter for extrusion is about 0.25 to 0.40 mm, the melt has to be filtered in a filter pack before extrusion through the spinneret to remove contaminants that can block the capillary holes. Therefore, it is necessary to deliver the melt at high pressure upstream of the spinneret, which is achieved by the use of gear pumps. The "spin block," thus, consists of the pump, the filter pack, and the spinneret. The extruder delivers the melt to a number of spin blocks through a melt distribution system. The design of the melt distribution system, the filter pack, and the spinneret is based on the melt rheology and thermal degradation characteristics of the polymer.

The filaments extruded through the spinneret are quenched either by air flow or by a water quench bath. The PET, nylon, and PP fibers are melt spun in a vertical chimney with air quench, whereas monofilaments and tapes generally involve horizontal extrusion lines with quenching of the extrudate in a water bath. The spun filaments are drawn concurrently or in a subsequent operation to orient the polymer molecules along the fiber-axis for obtaining high strength and modulus. The melt spinning and drawing process variables have a major influence on the mechanical behavior of the filaments and fibers. The drawn filaments are finally annealed in a high-temperature oven under tension to lock in the orientation through controlled crystallization. The drawing and annealing operations influence the crystalline morphology of the fibers and thereby their dyeing characteristics.

11.3.2 Solid-State Processing of Polymers

A number of polymer processing operations involve deformation of the polymer in the solid state at temperatures above its glass-transition temperature and below its melting point. There are two broad categories, namely, powder processing and post-extrusion processing.

Certain polymers such as polytetra-fluoroethylene (PTFE) and ultrahigh-molecular-weight polyethylene (UHMWPE) cannot be processed by the melt processing techniques because of their very high melt viscosities (of the order of 10,000 P) at the processing temperatures. These polymers exhibit high melt viscosities because of their extremely high molecular weights in the range of a few millions. Melt processing of such polymers would require very high processing temperatures for lowering their melt viscosity, which would then lead to thermal degradation of the polymer. Therefore, these polymers are processed by powder metallurgical techniques, such as solid-state compaction/sintering and hydrostatic extrusion.

The mechanical properties of a variety of extruded products such as films and fibers can be enhanced significantly in subsequent solid-state-drawing by orienting the polymer molecules along the direction of deformation. Also, extruded sheets of polymers can be transformed into different shapes by cold-drawing the sheets above the T_g but below the melting or softening temperature, in a process comparable to the metal sheet stamping process.

The different solid processing techniques are summarized as follows:

- Powder Processing
  Powder compaction/sintering
  Hydrostatic extrusion
- Drawing
  Film drawing
  Fiber drawing and annealing
- Sheet Forming
  Thermoforming
  Hot press forming of reinforced sheets

Powder Processing

The powder processing of polymers involves two steps, namely, powder compaction to make the preform or billet, followed by programmed sintering of the billet.

The compaction operation is carried out at room temperature in a conventional hydraulic press to produce standard preform shapes such as solid cylinders, annular cross-section cylinders, rectangular pellets, etc. The mold is filled with a weighed quantity of the powder and is then transferred to the press. The compacting plunger rod is then pressed *gradually* by downward movement of the upper platen until the desired compaction pressure is reached. After holding the pressure for a predetermined time (2–5 min), the pressure is slowly released. The compaction pressure is typically in the range 250–750 kg/cm^2. The maximum compaction pressure and the pressure cycle timings (compaction/holding release) are governed by both the part geometry the powder characteristics such as the particle size distribution, filler type and content, polymer compressive strength, etc. For billet L/D ratios above 1.5, it is desirable to design the mold for double compaction with plungers at the bottom and top of the mold to ensure uniform pressure transmission along the height of the billet.

The molded billet is then transferred to a sintering oven. The sintering cycle involves heating, holding at high temperatures, and cooling. A typical cycle for PTFE billets is shown in Figure 11.7. The major objective of the sintering step is to achieve adequate interparticle adhesion through diffusion bonding. Thus, the maximum plateau temperature is set above the polymer melting or softening point. Also, prior to reaching the maximum temperature, there is a preheating

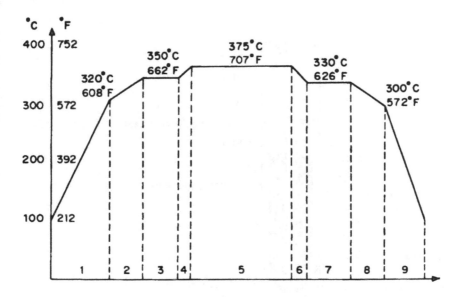

SINTERING CYCLE

Cycle	Ø in mm	1	2	3	4	5	6	7	8	9	Total
A	60	←——— 5h ———→				←—— 4h ——→		←——— 5h ———→			14h
B	100	3h	1 h 30	1h 30	1h	5h	1 h 30	1h 30	1h 30	3h	20h
C	120	4h	2 h	2 h	1h	6 h	1 h 30	2 h	2 h	4h	24h
D	150	6 h	3 h	3 h	1h	8 h	1 h 30	3 h	3 h	6h	34h
E	200	9h	6 h	6 h	1 h 30	14 h	2 h 30	6 h	6 h	9h	60h
F	150 x 350	10 h	3 h	3 h	2 h	12 h	2 h	3 h	3 h	10h	48h

Figure 11.7 Typical cycles for PTFE billets.

plateau around the temperature at onset of melting to ensure uniform temperature distribution across the part cross section. The heating is carried out gradually at 2–4 °C/min. Because the thermal conductivity of the polymer is low, it is necessary to provide adequate dwell time at the "plateau" temperatures to ensure uniform temperature throughout the cross section of the part. The dwell time parameters are thus governed mainly by the part geometry and the specific heat and thermal conductivity of the powder mix. The dwell time for the diffusion bonding plateau is to be defined with reference to the melt diffusivity of the

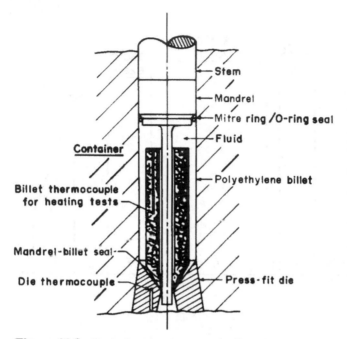

Figure 11.8 Typical equipment setup for hydrostatic extrusion.

polymer. The last stage of the sintering cycle involves slow cooling (0.5–1 °C/min) up to the T_g of the polymer followed by gradual cooling at normal cooling rates of 2–5 °C/min. The initial slow cooling is essential for minimizing residual stresses in the molded part and for achieving uniform crystallinity in the case of crystalline polymers such as PTFE and UHMWPE. The reader is referred to the review on powder processing of polymers by Bigg for further details [3].

The molded billet can be machined to make various components. Skived tapes of PTFE are obtained by slicing off a film of fixed thickness from the rotating annular cylindrical billet.

The cylindrical billet is also used for hydrostatic extrusion. In this process, the ductile material below its softening point but above its T_g is forced through a die by pressure transmitted through the hydrostatic fluid. Figure 11.8 shows a typical extrusion arrangement [4]. The low-viscosity fluid surrounding the billet lubricates the flow through the die, thereby reducing the extrusion pressure. The ratio of the billet area to the die area is termed the extrusion ratio, which is analogous to the draw ratio in tensile deformation. It determines the pressure drop, output rate, and properties of the extruded film or profile influenced by molecular orientation, as illustrated in Table 11.5 for hydrostatically extruded HDPE films.

Table 11.5 Effect of hydrostatic extrustion ratio on polymer properties.

Extrusion ratio	Tensile modulus (psi)	Elongation to break (%)	Melting temperature (°C)
1	200,000	600	132
3	400,000	22	133
13	2,500,000	10	137
35	10,000,000	3	140

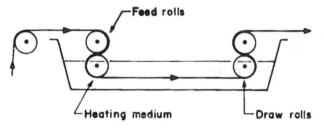

Figure 11.9 The drawing process.

Drawing

The drawing of cast films and melt-spun filaments of thermoplastics involves tensile deformation in the solid state. The drawing process is carried out continuously as illustrated schematically in Fig. 11.9. Tensile deformation of the film or filament "tow" (bundle or sliver of filaments) takes place because the speed of the draw rolls is higher than that of the feed rolls. The ratio of the speeds is termed the draw ratio. The drawing is carried out at a fixed temperature above the polymer T_g. The heating medium could be a hot water bath, a steam chamber, or a hot plate. In the drawing process, the polymer molecules are oriented along the draw direction leading to a significant improvement in the tensile strength and modulus of the fibers and films. The higher the draw ratio, the higher are the tensile properties. However, the maximum draw ratio is limited by the dynamic breakage of the filaments or film. The drawing temperature is a key process variable which is used for minimizing the draw force and yet increasing the orientation level.

In the processing of films and synthetic fibers of semicrystalline polymers such as HDPE, PP, PET, nylons, etc., the drawing step is followed by an annealing step. Annealing involves "setting" or "locking" the molecular orientation through crystallization in the solid state under tension. The drawn film or filaments are

STRAIGHT VACUUM FORMING

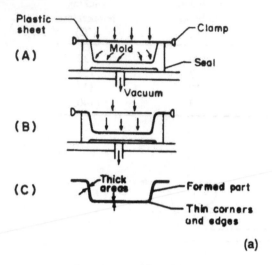

(a)

DRAPE FORMING

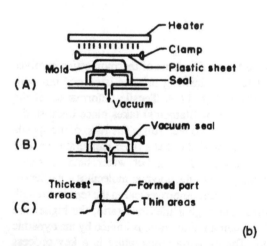

(b)

Figure 11.10 (a) Straight vacuum forming: The plastic sheet is clamped and heated. A vacuum beneath the sheet (A) then causes atmospheric pressure to push the sheet down into the mold. As the plastic contacts the mold (B), it cools. Areas of the sheet reaching the mold last are thinnest (C). (b) Drape forming: The plastic sheet is clamped and heated (A), then drawn over the mold either by pulling it over, or by forcing the mold into the sheet. A seal is created (B): vacuum applied beneath the mold forces the sheet over the

thus passed over a series of heated rolls to set in the orientation and control the shrinkage of the finished film or fiber product.

Sheet Forming

Thermoforming of unreinforced thermoplastic sheets is an economical process for converting flat sheets into three-dimensional shapes such as cups, shells, trays, etc. The process involves heating the plastic sheet above the T_g of the polymer to make it formable, followed by application of vacuum beneath the mold surface, thus forcing the sheet to conform to the shape contours of the male or female mold configuration. Depending on the mold configuration, the process is termed straight vacuum forming or drape forming as shown in Fig. 11.10. The thermoplastics readily amenable to thermoforming include PVC, ABS, HIPS, PP, and polycarbonate. The mold costs for the process are low and, therefore, it is an effective process for low- to medium-volume production of parts such as inner liners of refrigerator doors, canopies, small fishing boat hulls, kitchen sinks, etc.

The hot press forming or matched mold forming process is a compression molding process wherein the heated solid sheet is shaped by using a mold and a countermatching ram, as shown in Fig. 11.11 [5]. The process is increasingly being used for forming shapes from fiber-reinforced thermoplastic sheets because it allows the use of metal stamping presses with some modifications.

It may be noted that the sheet forming techniques cannot be used for very deep shaped articles due to the possibilities of extreme thickness variation in the part that could compromise its mechanical performance.

11.3.3 Solution Processing

Because the solution processing techniques involve the use of expensive and often toxic solvents, they are used only when melt or solid-state processing techniques are not possible. Solution processing is not very common and it is employed in exceptional cases as in the production of acrylic fibers, thin-film composite membranes, and aramid fibers.

Solution processing of polymers involves the following steps:

- Preparation of a homogeneous polymer solution, termed "Spin Dope"
- Extrusion of the viscous polymer solution
- Removal of the solvent in a coagulation medium (wet spinning) and/or by solvent evaporation (dry spinning)

---

male mold. The mold remains close to the original thickness of the sheet. Sidewalls are formed from the material draped between the top edges of the mold and the bottom seal area at the base. Final wall-thickness distribution is shown in (C).

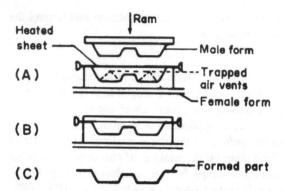

Figure 11.11 Matched mold forming: Matched molds of wood, metal, plaster, epoxy, etc., can be used to press the sheet into shape. The heated sheet may be clamped over the female die (A) or can be draped. As the mold closes, it forms the sheet (B). Mold vents allow trapped air to escape. Clearance between the mold force and cavity depends on tolerances required in the part. Excellent reproduction of mold detail and dimensional accuracy can be obtained, including lettering and grained surfaces. Material distribution of the formed part (C) will depend upon the shapes of the two forms.

Thus, the production of synthetic fibers from polyacrylonitrile (PAN) involves preparation of the spin dope of PAN in a solvent such as dimethyl formamide (DMF) followed by wet spinning or dry spinning. In wet spinning, the spin dope, typically containing 25–35% solids, is extruded through a spinneret horizontally into a bath containing the coagulating medium, which is generally a DMF/water mixture. The structure of the fiber is governed by the rate of precipitation of the polymer from the solution and is controlled by manipulating the amount of the nonsolvent in the spinning bath and the temperature. The as-spun filaments containing residual solvent (20–30%) are subsequently drawn and annealed for imparting orientation; the remainder of the solvent is removed in these steps. In the dry spinning process, the spin dope is extruded through a spinneret into a vertical heated chimney with the countercurrent flow of hot air. The solvent is, thus, removed by evaporation. The fiber structure is governed by the consolidation process, controlled by process parameters such as the wind-up speed, hot air temperature, and flow rate, chimney temperature, etc. The dry-spun filaments are also subsequently drawn and annealed. The dry spinning process allows considerably higher spinning speeds than the wet spinning process.

The solution spinning processes are required to be used for acrylic fibers because the melting point of polyacrylonitrile is very high and it starts degrading before it can be melted.

In the membrane casting process, the polymer solution is extruded through a film die and cast onto a heated substrate. Subsequent solvent removal and structure consolidation may be carried out in a coagulation bath or in an oven.

11.3.4 Reactive Processing

The processing of thermosetting polymers involves a chemical reaction that results in the formation of a three-dimensional network structure:

Liquid or solid reactants $\xrightarrow{\text{heat, pressure, catalyst}}$ solid infusible product

Such processing methods wherein the shaping operation is accompanied by a chemical reaction are categorized under "Reactive Processing." The examples of reactive processing techniques are given as follows:

- Liquid casting of epoxies for electrical insulation products
- Hand lay-up and spray-up for fiber-reinforced composite products
- Resin transfer molding for glass-fiber-reinforced unsaturated polyester products
- Compression molding of sheet molding compounds (SMC)
- Filament winding for fiber-reinforced plastic (FRP) pipes
- Pultrusion for FRP profiles
- Reaction injection molding of polyurethane foam products

In recent years, reactive processing has been extended to thermoplastic materials for producing specialty graft copolymers in an extruder reactor. The reactive extrusion technique is increasingly being used for compatibilization of immiscible polymer blends, dynamic vulcanization of thermoplastic/elastomer blends, etc.

The processing time in the reactive processing methods is controlled mainly by the kinetics of curing of the thermosetting polymers, which can be manipulated by temperature and the use of catalysts. Thus, the reactive processing methods for composite materials exhibit processing cycle times and, hence, productivity over a broad range; for example, the contact molding techniques such as hand lay-up and spray-up involve manual operations with long cycle times of the order of a few hours for a product like a bathtub. In hand lay-up, the fiber mat is manually spread over a wooden or epoxy mold surface conforming to the product shape and the resin–catalyst mixture is then spread over it using rollers and brushes. The curing takes place at room temperature. Once the resin is "gelled" (that is, it ceases to flow), the part may be demolded and allowed to cure further for a period of up to 24 hr. The same FRP products can be manufactured by using resin transfer molding (RTM), wherein the reinforcing fiber mats in the form of a preform are encased in a closed mold and then the liquid resin mixed with the catalyst and promoter is pumped in. The curing reaction takes place in the mold cavity and can be accelerated by heating the mold. The cycle time could, thus, be as low as 10–15 min.

Compression molding of SMC and DMC (dough molding compounds) represents a high-productivity reactive processing technique for making large-volume composite products, with cycle times of the order of a few minutes (2–5 min). The processing involves curing of the fiber–resin mix under high pressure (100 psi) and high temperature (100 °C). Because of the severe processing conditions, the dies for compression molding are required to be fabricated from hardened alloy steels and, therefore, these are considerably more expensive. Compression molding also requires investment in hydraulic presses.

The hand lay-up/spray-up, RTM, and compression molding are batch processing techniques for composite materials. The productivity of the process increases as one goes from the hand lay-up process to compression molding. However, the capital investment in machinery and tools (dies) is also more for the high-productivity automatic molding process. The selection of the process would, thus, depend on the production volume, size, and shape of the product and curing rate of the resin. For example, large-volume, small-size products such as switch-gear components, automotive components like distributor caps, etc., are produced by compression molding of SMC and DMC. Whereas large-size FRP products required in relatively smaller quantities, such as bathtub assembly, fishing boats, truck cabins, etc., are generally produced by the hand lay-up process or RTM.

Filament winding represents a semicontinuous processing method for composite materials. In filament winding, continuous rovings (filament bundles) of the reinforcement (glass, carbon, or aramid) pass through a resin–catalyst mix and the resin-impregnated rovings are then wound onto a rotating cylindrical mandrel, along the length of the mandrel by moving the roving feed point laterally in a reciprocating manner at a fixed speed. The relative speeds of rotation of the mandrel and translation of the roving feed determine the angle of the reinforcement and thereby the mechanical properties (for example, limiting hoop stress) of the composite pipe. Noncylindrical product shapes such as conduits and ducts with rectangular or triangular crosssection can also be produced by filament winding. Corrosion-resistant FRP pipes, storage tanks, columns, etc., for the chemical process industries are produced by this technique. Because curing of the resin generally takes place at room temperature in filament winding, the resin mix containing the catalyst/promoter has to be so designed that adequate processing time is allowed for completion of the winding operation until the desired wall thickness of the pipe is obtained. Subsequent curing of the resin can be carried out at room temperature; the curing may also be accelerated by the use of hot air blowers. However, it is important to ensure a uniform rate of curing along the length and circumference of the pipe to prevent warpage due to residual stresses.

The pultrusion process is a continuous process used for producing standard shape profiles such as tubes, C-sections, L-sections, E-sections, etc., from FRP materials. In this process, the continuous roving reinforcement, impregnated with the resin–catalyst mix, is fed into a heated die having a cross section of the desired profile. The temperature profile along the die length is set to achieve complete curing of the resin at the downstream end of the die. The solidified profile is pulled continuously downstream by a puller and then cut into standard lengths by an on-line cutter. A thorough understanding of the curing kinetics of the resin and the effect of temperature, catalyst concentration, etc., on the curing rate is required for designing the pultrusion die and also for optimizing the temperature profile.

In both filament winding and pultrusion, the flow behavior of the resin catalyst mix is required to be controlled to avoid dripping of the resin from the impregnated roving and to ensure good uniform wetting. This is achieved by using thixotropic additives.

It is clear from the preceding discussion that the important scientific database required for effective design, optimization, and control of reactive processing of composite materials include kinetics of curing as affected by process parameters such as temperature and catalyst/promoter concentration, and change in resin viscosity with degree of cure (termed "chemorheology"). The most common thermosetting polymers subjected to reactive processing are unsaturated polyesters, epoxies, phenolics, and polyimides.

The production of flexible and rigid foam products and reinforced microcellular elastomeric products of polyurethanes also involves reactive processing. The processes are, therefore, termed reaction injection molding (RIM) or reinforced reaction injection molding (RRIM). The main distinguishing feature of polyurethane reactive processing relative to the composite processing discussed earlier is the extremely short reaction time of the order of a few seconds. The RIM equipment is, thus, considerably more sophisticated. In RIM processing, the polyol and isocyanate components of the PU system are pumped from separate storage tanks, maintained at preset temperatures, to a mixing chamber. The thorough mixing of the two streams is achieved either by high-speed mechanical rotors or by impingement of the high-pressure jets of the two streams. The mixture is then poured into the closed mold cavity. Because the time available before onset of foaming is typically of the order of 5–10 sec, the mixing and pouring operations have to be completed in this short period. The chemical reaction between the polyol (resin) and isocyanate (hardener) components takes place in the mold cavity and foaming is generally achieved by evaporation of the blowing agent, premixed in the polyol component, as a result of the exothermic reaction. The most commonly used blowing agent is Freon R-11. The foam density is controlled by the amount of the blowing agent, the resin/hardener ratio, the efficiency of mixing, and the mixture temperature. In a number of foam products, formation of an integral high-density skin is desirable, and this can be achieved by controlling the surface temperature of the mold cavity. Besides the blowing agent, surfactants are often added to the PU system for obtaining uniform cell size in the foam.

The various products manufactured by using RIM processes include automotive bumpers of microcellular elastomers, flexible foam seats, rigid foam insulation panels, etc.

11.4 Processibility of Polymers

The processing of polymers requires knowledge of the thermal behavior and flow characteristics of the polymer in the case of thermoplastic materials, and knowledge about the curing kinetics and gelation behavior of the thermosetting resins.

11.4.1 Thermal Characteristics

The thermal characteristics of thermoplastics is specified in terms of the degradation temperature, melting temperature, and glass-transition temperature.

For every polymer, there is a maximum temperature beyond which chain scission takes place leading to depolymerization. Thus, the processing temperature has to be kept below the degradation temperature, T_d. The temperature of onset of thermal degradation and the rate of degradation are also affected by the environment. Generally, the degradation is facilitated in the presence of air or oxygen. The data on thermal stability of polymers are obtained by the technique of thermogravimetric analysis (TGA). The instrumental analysis involves monitoring the weight loss as a function of temperature. It is carried out in nitrogen atmosphere to determine the thermal degradation temperature, whereas the analysis in air gives the oxidative degradation temperature. The data on the thermal degradation temperatures of polypropylene, polystyrene, and poly(ethylene terephthalate) are given in Table 11.6 [6]. The data clearly indicate that the degradation temperature is lower in air atmosphere relative to the T_d in nitrogen atmosphere. The typical TGA curves are shown in Figs. 11.12 and 11.13.

TABLE 11.6 TGA data on thermal stability of thermoplastics.

Polymer	Grade	Pure thermal degradation temp. (N_2 atm.), °C	Oxidative degradation temp. (Air atm.), °C
Polypropylene	Hercules PC-973	255	238
Polystyrene	Dow Styron 666U	258	250
Poly(ethylene terephthalate)	Goodyear VFR-4704A	350	310

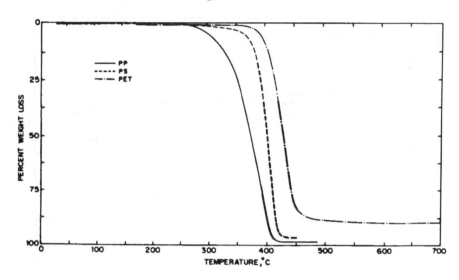

Figure 11.12 Pure thermal degradation of PP, PS, and PET by TGA.

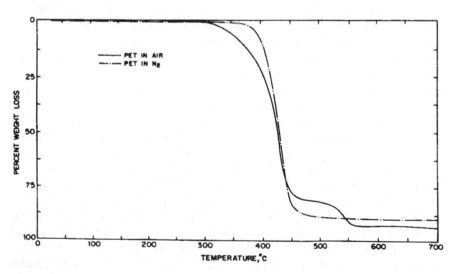

Figure 11.13 TGA data for PET in air and nitrogen atmospheres.

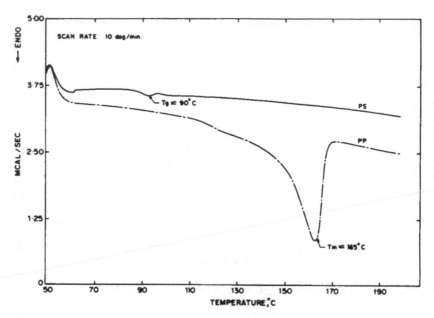

Figure 11.14 DSC heating scans of polypropylene (---) and polystyrene (−).

The polymer melt is susceptible to purely thermal degradation during continuous flow operation when it comes in contact with overheated internal surfaces of the extruder barrel, screw, molds, and transfer lines. Oxidative degradation is encountered when the polymer melt is extruded through the die into an open air atmosphere. During the initial heating and final cooling cycles of a batch operation, the melt trapped in mold cavities, screw and barrel surfaces, filter packs, etc., is also subjected to oxidative degradation. Therefore, the maximum temperature at any point in the process must be kept well below (approximately 10 °C) the pure thermal degradation temperature of the polymer.

The processing temperature has to be above the melting temperature of the semicrystalline polymer (T_m) or the softening temperature of the amorphous polymer. The melting temperature is determined by using the technique of differential scanning calorimetry (DSC). Typical DSC heating scans for a crystalline polymer (polypropylene) and an amorphous polymer (polystyrene) are shown in Fig. 11.14. A clear melting peak is observed for the crystalline polypropylene (165 °C) but not for the amorphous polystyrene. Thus, in the case of amorphous polymers, a softening temperature is determined as the temperature at which the solid polymer sample loses its mechanical integrity and starts flowing. The softening temperature of polystyrene is about 170 °C. The melting and softening points of a few of the common thermoplastics are summarized in Table 11.7.

Table 11.7 Glass-transition temperatures and melting/softening temperatures of thermoplastics.

Polymer	Type	T_g (°C)	T_m (°C)	Softening temp. °C
HDPE	Crystalline	−130	132	—
PP	Crystalline	−10	165	—
Polyacetals	Crystalline			—
PS	Amorphous	95	—	170
PMMA	Amorphous	90	—	165
Nylon 6	Crystalline	35	225	—
PBT	Crystalline	34	230	—
Nylon 66	Crystalline	50	260	—
PET	Crystalline	70	255	—
Polycarbonate	Amorphous	140	—	270
PPS	Crystalline	88	290	—
PTFE	Crystalline	125	345	—

A polymer is susceptible to thermal degradation even at the processing temperatures below T_d if the exposure time is long. The allowable residence time at a preset processing temperature prior to onset of degradation can be obtained by carrying out isothermal TGA of the polymer: The data for polypropylene at 190, 200, and 210 °C are summarized in Table 11.8. It is observed that little thermal degradation of PP occurs in nitrogen atmosphere even at 210 °C up to an exposure time of 60 min. However, in air atmosphere, weight loss of up to 5% is observed within 5 min at 210 °C, 7 min at 200 °C, and 10 min at 190 °C. These data are useful for optimization of the process and also for product quality control.

The glass-transition temperature of the polymer (T_g) is a relevant thermal parameter for solid-state forming processes. It is the temperature above which the solid polymer can undergo plastic deformation. The T_g of a polymer can be determined by DSC or by dynamic mechanical analysis (DMA). The T_g values of a few common thermoplastics are given in Table 11.7.

11.4.2 Flow Behavior

The response of polymeric fluids to deformation is significantly different from that of the common Newtonian fluids because of their long-chain "string-like" structure and polydispersity (molecular weight distribution) resulting in entanglements and strong intermolecular forces.

Conventional Newtonian fluids exhibit a constant viscosity not dependent on the level of applied stress or the resulting velocity gradient; the viscosity depends only on temperature and pressure as per the following equation:

Table 11.8 Isothermal TGA data on polypropylene grade: Hercules PC-973.

Time (min)	% weight loss		% weight loss		% weight loss	
	Air 190 °C	N_2 190 °C	Air 200 °C	N_2 200 °C	Air 210 °C	N_2 210 °C
5	2.50	1.38	3.75	1.0	5.0	0
10	5.00	1.38	8.13	1.13	10.63	0
15	8.75	1.38	13.13	1.13	16.25	0.125
20	12.50	1.38	17.5	1.25	21.88	0.125
25	16.25	1.38	22.5	1.25	27.5	0.25
30	20.00	1.5	26.88	1.25	33.75	0.38
40	28.13	1.63	35.63	1.38	44.38	0.63
50	36.25	1.88	45.00	1.5	54.38	0.75
60	43.75	2.0	51.88	1.63	58.75	1.13
Sample size (mg)	8.6	7.4	10.5	8.7	8.6	11.4

$$\mu(T,P) = \mu_0 \exp\left[\frac{\Delta E}{R} \frac{T_0 - T}{T_o T} \right] \exp[\beta(P - P_0)] \qquad (11.1)$$

where μ_0 is the viscosity at the reference temperature T_0 and pressure P_0, ΔE is the activation energy for flow, R is the gas constant, and β is a material property (m^2/N).

Newton's law describes the response of common fluids to applied stress, stating that the applied stress, τ, is directly proportional to the resulting rate of strain, $\dot{\gamma}$. The above equation is a linear equation because the "viscosity" (proportionality constant) is constant. However, in the case of polymeric fluids, the viscosity is dependent on the rate of deformation or the stress.

The flow fields encountered in processing are categorized as "shear flows," "extensional flows," or combinations of the two (Fig. 11.15). In shear flows, the velocity gradient is perpendicular to the velocity vector, whereas in extensional flows the velocity vector and the velocity gradient are parallel. The predominant flow inside extrusion dies and spinneret capillaries is shear flow, whereas the flow in spinning a molten threadline is extensional. Extensional flow is also relevant in blown film extrusion. The analysis of the flow fields during the "shaping" operation is important because of their physical effects on the "conformations" or "spatial arrangement" of macromolecules.

Polymer melts and solutions generally exhibit a shear thinning or pseudo-plastic behavior, meaning that the viscosity or the resistance of the fluid to deformation

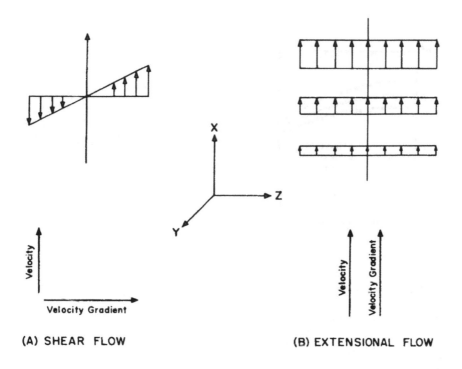

(A) SHEAR FLOW   (B) EXTENSIONAL FLOW

(Arrows represent velocity vectors)

Figure 11.15 Flow fields in polymer processing (arrows represent velocity vectors).

decreases with increasing stress. Phenomenologically, this could be explained in terms of a decrease in the "entanglement density" due to the "sliding" motion of adjacent macromolecules, under the rotational deformation in shear flow. This physical description would also imply greater difficulty in affecting disentanglement in an extensional flow field, wherein the extensional deformation may lead to "knotting" at entanglement points. Some amount of chain extension and orientation would take place in the shear flow also, particularly for the chain segments between two entanglement points. The experimentally observed variation of the shear viscosity and extensional viscosity are shown in Fig. 11.16 [7]. Qualitatively the trends are clear:

At low shear rates, the shear viscosity is independent of the shear rate; that is, the fluid exhibits Newtonian flow behavior.

Beyond a critical value of shear rate (dependent on temperatures and the material parameters), the viscosity decreases with increasing shear rate in a nearly

(a)

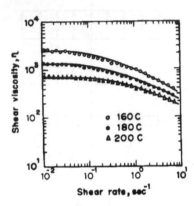

(b)

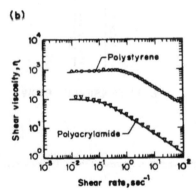

(c)

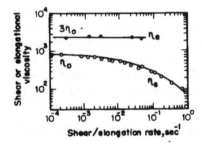

Figure 11.16 Flow behavior of polymeric fluids. (a) Shear viscosity of LDPE melt; (b) shear viscosity of polymer solutions; (c) shear and elongational viscosity of isobutylene-isoprene.

straightline manner on *double-logarithmic coordinates*, signifying a nonlinear power law relationship between viscosity and shear rate:

$$\eta_s = \dot{\gamma}^m \tag{11.2}$$

The extensional viscosity η_e is much higher than the shear viscosity η_s and is not dependent on the extension rate (for the case of the rubber).

Besides differences in the viscous behavior, the polymeric fluids exhibit partial elastic response. The classical experiment illustrating the different flow response of polymeric fluids versus Newtonian fluids involves stirring the fluid with a rod at the center. What is observed is that the Newtonian fluid forms a "concave trough" at the center, whereas the polymeric fluid starts "climbing up" the rod. This inward motion is against the centrifugal forces. This phenomenon termed the "Weissenberg effect" is observed even at low shear rates. The physical explanation is that the polymer molecules become oriented during the annular flow, and in their tendency to return to the random coil state, they exert a hoop stress component on the adjacent fluid layer toward the rod. The processing observations of "die swell" and "melt fracture" of the extrudate are manifestations of this partial elastic response.

Another characteristic flow behavior of polymer melts and solutions is their sluggish stress relaxation response on removal of the imposed stress or strain. This indicates that the molecular conformations achieved due to the effect of flow fields "retract" (partially) to the stable random coil conformations in a "finite time" termed the relaxation time mainly dependent on the temperature. The relative ratio of the "processing time" and the "relaxation time" would thus govern the "residual stresses" in the shaped article, thus influencing both the morphology and properties.

The knowledge of the flow behavior of polymer melts and solutions is essential for design of processing equipment and also in analyzing product quality problems and performance failures.

For further details on the rheology of polymeric fluids, the reader is referred to standard texts on the subject [7].

11.4.3 Polymer Structure and Processibility

The thermal degradation characteristics of a polymer are mainly governed by its chemical structure. In the case of copolymers, the comonomer composition and sequence length distribution (CCD and SLD) may affect the thermal stability to some extent.

The melting behavior of semicrystalline polymers is influenced by the morphology in the solid state, defined in terms of the spherulite and lamellar size distribution. However, it is only the melting temperature range that is generally affected. The melting peak temperature is affected by the type of crystalline

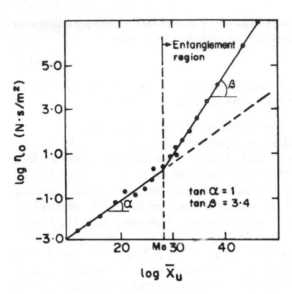

η_o - Zero shear viscosity

$\overline{X}_u$ - Weight average chain length

Me - Entanglement molecular mass

Figure 11.17 Dependence of viscosity on molecular weight.

morphology. The "extended chain" crystalline morphology is known to exhibit a higher melting temperature than that of the "folded chain" crystals. However, even for polyethylene, the increase is only about 7–8°C.

In general, beyond a critical molecular weight, the thermal degradation and melting behavior and the glass-transition temperature of the polymer are not affected much by changes in the "molecular architecture" (molecular weight distribution, copolymer sequence distribution, etc.).

The processibility parameters that are most sensitive to the polymer structure parameter, such as MWD, are the viscosity and curing behavior (for thermosets). The viscosity of liquid resins, polymer melts, and solutions is dependent on both the molecular weight and the MWD. Figure 11.17 illustrates the variation of the zero shear viscosity, η_0 (that is, the viscosity in the limit of zero shear rate) with the molecular weight [8]. It is seen that η is directly proportional to the molecular weight up to a critical molecular weight, beyond which the dependence of η on M is much stronger ($\eta_s \propto M^{3.5}$). The occurrence of the critical molecular weight is attributed to the onset of entanglement formation beyond a particular chain

Table 11.9 Effect of MWD on processibility of high-density polyethylene.

Resin	Melt index	MWD	Melt temperature at 10 rpm (°C)	Critical shear rate (sec^{-1})
A	0.8	Narrow	192	2000
B	0.8	Broad	184	4000

length. The existence of chain entanglements would enhance the dependence of viscosity on the molecular weight.

The molecular weight distribution has a significant influence on the shear viscosity versus shear rate curve and, hence, on processibility. As far as processing is concerned, knowledge of the complete flow curve is essential because both the high and low shear flow characteristics are critical depending on the processing technique employed. High melt viscosity at low shear is needed in processes such as blow-molding, where excessive drawdown can be a serious problem. On the other hand, in injection molding, involving high shear rates in mold filling, low melt viscosity at high shear is needed to minimize the amount of heat generated by viscous friction and to attain high productivity before melt flow instability occurs. Table 11.9 illustrates that at the same melt index (MI), the resin B with a lower high shear viscosity (broad MWD) runs about 8 °C cooler and exhibits a higher critical shear rate at the onset of flow instability than the resin A with a higher high shear viscosity (narrow MWD) [9].

Another example of the sensitivity of flow and processing behavior to MWD is the effect of the content of high-molecular-weight (HMW) species on the blown film processibility of LDPE. The GPC (gel permeation chromatograph) data on the MWD of two LDPE resins (A and B, and their corresponding curves are shown in Figs. 11.18 and 11.19. Resin A had a density of 0.927 and a MI of 0.51; corresponding values for resin B are 0.929 and 0.54. The GPC data indicate that resin A has a greater content of HMW species probably due to greater extent of long-chain branching. Although they may comprise only a small fraction of the total molecular population, the longer branched molecules influence such processing characteristics as die swell, drawdown, and neck-in. Melt flow properties are also affected by the presence of the HMW species, but differences are detectable only at very low shear rates; the melt index values fail to differentiate between the two resins. It is expected that resin A would exhibit greater die swell, poorer drawdown in tubular blown film extrusion, and lower neck-in in extrusion coating. Also, resin A would yield a film with higher impact strength and haze levels.

In the case of synthesis of thermosetting polymers such as Novolac phenolic resins, variation in the molecular structure in terms of molecular weight distribution

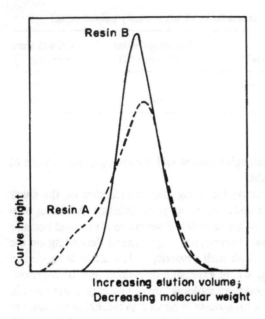

Figure 11.18 Elution curves for two LDPE resins differing greatly in content of HMW species.

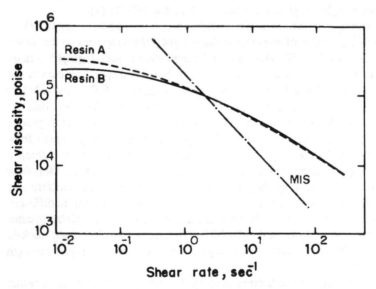

Figure 11.19 Flow curves for LDPE. MIS = melt index stress with 2.16 kg weight.

Table 11.10 Effect of structure variation on flow and curing characteristics of Novolac phenolic resins.

Sample code	A	B	C
Solution viscosity[a] (cps)	160	220	270
Norton flow[b] (mm)	70	46	32
Melting point (°C)	68	69	72
Molecular weight[c] (Mn)	625	646	661
MWD[d]	Unimodal with shoulder	Weak bimodal	Bimodal

[a]Viscosity of 60% solution of solid resin in methanol.
[b]Flow length of a pellet of resin with 6% hexamethylene tetramine at 125°C.
[c]Vapor pressure osmometry.
[d]High-pressure liquid chromatography.

and linearity/branching of the polymer can affect both the flow characteristics and curing behavior of the resin, as illustrated in Table 11.10. Referring to Table 11.10, although the three samples of the Novolac resin were produced in the same reactor using identical stoichiometry, these vary significantly in their solution viscosity (60% solid resin in methanol) and curing behavior (Norton flow parameter). Resin A with a unimodal MWD exhibited the lowest viscosity and the slowest cure (high Norton flow length), whereas resin C with a bimodal MWD and higher molecular weight manifested the highest viscosity and the fastest cure. This variation in structure and processibility is the result of lack of proper process control in the polymerization reactor.

The above examples clearly illustrate the influence of molecular architecture (mainly MWD) on the processibility. The control of polymerization reactor conditions is, therefore, critical to ensure uniform structure and, hence, consistent processibility of the polymer.

11.5 Structure Development in Polymer Processing

As a result of their long, chain-like structure constituting a series of strong covalent bonds, polymers can exhibit a variety of macroscopic conformations in the liquid state (solution or melt), depending on the stearic, thermal, and flow-induced stress fields. The conformations would include a random coil, folded chain, or an extended chain conformation on a long-range (length scale) basis, and also random, zigzag, or helical conformations, when viewed over a shorter length scale of a few bonded monomeric chain segments.

The structure development in polymer processing concerns the effect of the thermal and mechanical environment on the intermolecular and intramolecular

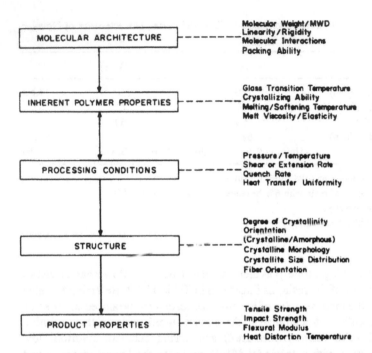

Figure 11.20 Processing–structure–property relations in polymers.

changes in the polymer chain conformations. The manner in which the macromolecules are packed and oriented with respect to each other at the end of processing would determine the properties of the shaped articles. The structure is the integrated result of a number of competing physical processes acting on the polymer molecules during processing, such as flow-induced molecular orientation, relaxation, cooling, and crystallization. The interdependence of the properties, structure, and processing parameters in polymeric materials is schematically illustrated in the form of a block diagram in Fig. 11.20. Therefore, it is critical to control the process parameters to ensure consistent morphology and, hence, properties of the processed article.

The key parameters and features that define the morphology or physical structure of the polymer in the solid state are discussed below. Unlike metals, semicrystalline polymers exhibit a two-phase structure in the solid state, consisting of distinct crystalline and amorphous domains. The volume fraction of the material constituting the crystalline phase is termed the "degree of crystallinity." The basic structural unit in the crystalline domains is the "folded chain lamella." The lamellar structure is formed by the chain folding of a number of polymer chains. Any single polymer chain can enter and leave the lamella. Thus, a fraction of

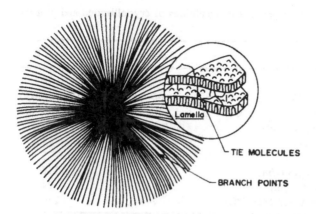

TIE MOLECULES

BRANCH POINTS

Figure 11.21 Schematic representation of spherulitic structure.

the segments of any chain forms the crystalline phase, whereas the remaining segments constitute the amorphous phase. This feature gives rise to the interconnections between the two phases essential for stress transfer during deformation. The lamellae, in turn, pack into a spherical geometry giving rise to "spherulites" which represent a supermolecular structural unit in polymers crystallized from melts (Fig. 11.21). The nature and extent of the "interconnections" or "tie molecules" determine the stress–strain behavior of the processed article.

The stress fields and orientation effects experienced by the polymer molecules in the molten state influence both the crystallization kinetics and the nature of the supermolecular morphology, exhibited in the form of either spherulites, row nucleated axialites, or "shish-kebab" structures.

Thus, the parameters defining the "structure" in the solid state of a semicrystalline polymer include degree of crystallinity, lamellar or crystallite thickness, type and size of the supermolecular morphology (for example, spherulite size distribution), the nature and extent of "tie molecules," and molecular orientation.

In the case of injection molding of semicrystalline engineering plastics such as nylons, thermoplastic polyesters, polyphenylene sulfide (PPS), polyether-ether-ketone (PEEK), etc., crystallization of the polymer takes place during the molding cycle. The rate of cooling in the mold has considerable influence on the degree of crystallinity and the resulting morphology, which, in turn, affects the mechanical properties and dimensional stability of the molded part. Therefore, it is necessary to optimize and control the injection molding parameters such as the melt temperature, mold temperature, and demolding time for a given product in the context of the relation between the competing rate processes of crystallization and cooling.

Table 11.11 Effect of mold temperature on conditions of crystallization and properties of PPS. (From Ref. 11.)

	Set A	Set B
Mold temperature, T_w (°C)	135	38
Range of quench rates, $q°$ (°C/min)	60–660	1020–1620
Crystallization temperature range, ΔT_c (°C)	158–200	125–150
Required crystallization time, t_c (sec)	9.5–13	11–33
"Dwell time" over ΔT_c, t_d (sec)	9	1.5
Degree of crystallinity (%)	50	5
Unnotched Izod impact strength (J/m)	175	312
Notched Izod impact strength (J/m)	58	74
Heat distortion temperature at 1.8 Mpa (°C)	243	230

The primary consideration in specifying the molding conditions is that the crystallization of the polymer should be complete throughout the cross section of the molded part before demolding. In the case of incomplete and nonuniform crystalline morphology, shrinkage, and warpage of the molded part would result, particularly when the molded part is exposed to temperatures above the glass-transition temperature of the polymer. It is, therefore, critical to manipulate the cooling conditions in such a way that the polymer spends adequate time in the mold at temperatures over which the rate of crystallization of the polymer is fast, so that complete crystallization is ensured before demolding. Thus, the inherent crystallization kinetics of the polymer is the limiting factor in specifying the mold cycle time, although a short mold cycle is preferred to increase productivity.

The effect of processing conditions on the degree of crystallinity and on properties of injection-molded PPS is illustrated in Table 11.11 [11]. It is clear that the molding conditions in set A, wherein the mold temperature was held at 135 °C, are conducive to developing equilibrium crystallinity in the molded part. For set A, under the relatively slow quench rates, the crystallization temperatures (5–200 °C) lie within the range of fast crystallization for PPS (142–205 °C); the crystallization temperatures are obtained from crystallization kinetics studies. Once crystallization is induced, the d well time (t_d) spent by the polymer over the relevant temperature range is comparable to the crystallization time, t_c, in set A. On the other hand, for set B, the range of crystallization temperature is narrow and the dwell time of the polymer over this range is significantly shorter than the required crystallization time. Thus, little crystallinity development would take place under the molding conditions of set B, wherein the mold temperature was held at 38 °C. The conclusions are consistent with experimental results of Brady [12] who has reported a crystallinity index of 50% for PPS molded at a mold

Introduction to Polymer Processing 341

Table 11.12 Effect of molding conditions on mechanical properties of polypropylene. (From Ref. 13.)

	Batch no.		
	1	2	3
Melt temperature, T_0 (°C)	230	230	230
Mold temperature, T_w (°C)	12	45	105
Degree of crystallinity (%)	43	46	48
Spherulite size (μm)	10	30	60
Skin thickness, t (μm)	325	310	290
Tensile strength (MPa)	34	35.8	36.9
Elongation at break (%)	600	600	30
Tensile modulus (MPa)	1	1.8 / 9	1.9 / 2
Impact transition temperature (°C)	30	26	23

temperature of 135 °C and a crystallinity index of 5% for molding at a mold temperature of 38 °C. This lack of crystallinity in sample B is the reason for the lower heat distortion temperature and higher impact strength.

The processing conditions affect not only the degree of crystallinity but also the spherulite size distribution; the properties of the molded part are influenced significantly by the spherulite size distribution, even if the degree of crystallinity values may be comparable. This is illustrated by the data on polypropylene and polyacetals summarized in Tables 11.12 and 11.13 [13]. At a given melt temperature, a lower mold temperature increases the quench rate leading to lower crystallinity and smaller spherulites. Similarly, a low melt temperature close to the melting point of the polymer gives rise to a high density of nuclei on the mold surface leading to small spherulites. Small spherulite size improves impact resistance, whereas large spherulites could significantly lower the value of the strain to failure and, hence, the toughness. Thus, changes in the spherulitic size distribution effected through variations in processing conditions significantly influence product properties.

The above discussion underscores the need for accurate control of the process

Table 11.13 Effect of molding conditions on impact behavior of polyacetal. (From Ref. 13.)

	Batch no.		
	1	2	3
Melt temperature, T_0 (°C)	210	200	188
Mold temperature, T_w (°C)	65	96	130
Degree of crystallinity (%)	63	66	67
Spherulite size (μm)	160	60	20
Skin thickness, t (μm)	80	60	40
Tensile strength (MPa)	64	69	70
Tensile modulus (MPa)	3.07	3.48	3.54
Impact energy (Nm)	2.70	2.91	17.3

controlling the morphology and the product properties has been illustrated for nylon by Steinbach [17].

The other major effect of processing is the change in the orientation of the polymer molecules as a result of the mechanical deformation. The ability of polymers to be structured during processing is inherent to their anisotropic macromolecular structure, which is held together by strong covalent bonds along the chain axis and by weak, but many, secondary bonds between adjacent chains in the transverse direction. The molecular orientation is an important structural parameter influencing the properties of films and fibers. As illustrated in Table 11.14, the modulus of the most conventional polymer, HDPE, can be improved significantly by increasing the molecular orientation through innovative process variations.

In summary, the structure of the polymer in the solid state determines the properties of the processed article. The structure is formed in processing as a result of the interactions between the processing parameters, such as the cooling rate, stretching rate, etc., and the inherent polymer characteristics, such as relaxation rate, crystallization rate, etc. For crystallizing polymers, the rate of crystallization versus the rate of cooling would govern the degree of crystallinity, the spherulite size distribution and, hence, mechanical properties. For amorphous polymers, the relaxation time spectrum versus the processing time determines the extent of residual stresses and, hence, long-term stress crack resistance. The relative rates of flow deformation, relaxation to the random coil conformation and cooling determine the extent of orientation and, hence, the mechanical and optical properties. In view of the sensitivity of the solid-state structure and properties of the polymer to process parameters, it is essential to develop a framework of the control problem in polymer processing for instituting accurate control of the processing conditions for consistent desired product quality.

Table 11.14   Effect of molecular orientation on modulus of HDPE Fibers.

Process	Modulus (GPa)
Conventional drawing	1–7
Extrusion drawing [Ref. 18]	70
Cold ultra-drawing [Ref. 19]	68
Gel spinning/drawing [Ref. 20]	120
Theoretical limit	240

11.6 Identification and Control of Processing Variables

As discussed in the preceding section, the end-use properties of a polymer product, such as packaging film, synthetic fibers, molded article, or composite laminate are governed by the physical structure of the polymer. Because the objective of the processing operation is to produce a product with desired properties in a consistent manner, the control strategy would involve identification of the key end-use properties, the critical structural parameters that determine the end-use properties, and then the process variables to be controlled and manipulated for ensuring the desired "structure" consistently.

The product properties fall under different categories such as mechanical, thermal, electrical, optical, tribological, aesthetic, etc. Examples of mechanical properties include tensile and flexural strength, Young's modulus, impact strength, tear strength, etc. Heat distortion temperature specifies the thermal property limit of a product. Whereas dielectric strength, comparative track index (CTI), etc., represent "electrical" properties. The tribological properties are determined in terms of PV value obtained from the product of the pressure on and velocity of the moving part. The optical and aesthetic properties are often described in terms of gloss, clarity, haze, etc.

The hierarchy of the desirable product properties would vary depending on the end use or application. For example, the desired properties for fibers in textile applications (apparel fabrics) would include a balance of yield strength and dyeability, whereas for industrial fibers (ropes, twines, architectural fabrics) the major property requirements would be the highest possible strength and modulus. Similarly, for a PET film to be used in capacitors, the major property targets would be the dielectric strength and loss factor, whereas for application of the same film in food packaging, the critical property would be water vapor and oxygen permeability.

A majority of the *end-use properties* are not amenable to on-line measurements, with the possible exception of properties such as optical clarity and surface gloss. Therefore, it is necessary to identify *controlled parameters affecting the product quality*. In the case of thermoplastic products, the structure parameters governing the end-use properties are the degree of crystallinity, spherulite size distribution, molecular orientation, and residual stresses (for amorphous polymers). The key structure parameters governing end-use properties in thermoset composite products include the degree of cure, the uniformity of the degree of cure through the part cross section, and residual stresses, influenced mainly by the rate of curing. Again, most of these parameters are also not measurable on-line. A few of the parameters can be readily measured off-line. Thus, the degree of crystallinity can be determined from the density in a density gradient column or from the heat of fusion by DSC. The degree of cure may be found out by extraction in a suitable solvent. The molecular orientation of a fiber or film can be obtained by birefringence measurements using a polarizing optical microscope. However, determination of structure parameters such as spherulite size distribution and residual stresses is time-consuming and tedious, even in off-line measurements.

In view of the difficulties in on-line measurement of the structure parameters, certain *response parameters related to the dimensional uniformity of the product* are often used in polymer processing for monitoring the effectiveness of the process control. Thus, the film thickness variation across the width of a film is controlled within $+/-$ 5% for ensuring consistent process control and product quality. Similarly, the thickness variation along the circumference of a pipe, coefficient of variation of the filament diameters in a melt-spun filament bundle, and warpage of a molded flat plate are often used as tests for process control.

The *controlled variables specifying operating conditions* in polymer processing include the melt or stock temperature, the temperature profile on the different zones of the extruder, the take-up roller or wind-up speed for films or fibers, the injection pressure and velocity profiles in injection molding, and the temperature–time and pressure profiles in autoclave molding of composite laminates. The most commonly used *manipulated variables* are the screw rpm and die temperature in sheet extrusion; the quench air temperature and flow rate in blown film extrusion; the mold temperature, melt temperature and injection pressure profile in injection molding; the extrusion temperature, extrusion velocity and wind-up speed in melt spinning of synthetic fibers; and the curing temperature profile in composite processing.

Because the various polymer processing operations encompass a broad spectrum of materials, it may not be possible to evolve a generalized control strategy for polymer processing. However, it is possible to define a methodology that involves identifying the critical product properties, the structure parameters governing these properties, the process variables to which the structure is the most sensitive, and, hence, specifying the process control requirements. The specific controlled and manipulated variables will change depending on the process and the material type (thermoplastic versus thermoset) being processed.

References

1. N. J. Mills, *Plastics: Microstructure, Properties and Applications*, Edward Arnold Publishers, London, 1986.
2. "Moving Ahead with Rotational Molding," *Plastics Design Process.*, 24–29 (September 1979).
3. D. M. Bigg, "A Study of the Effect of Pressure, Time and Temperature on High Pressure Powder Molding," *Polym. Eng. Sci.*, *17*, 691–699 (1977).
4. D. M. Bigg, "Hydrostatic Extrusion of Semicrystalline Polymers for Superior Tensile Properties," *Plastics Design Process.*, 21–28 (February 1979).
5. C. A. Harper, "What you should know about Plastics Processing," *Chem. Eng.*, 100–114 (May 1976).
6. V. M. Nadkarni, private communications, 1988.
7. S. Middleman, *Fundamentals of Polymer Processing*, McGraw-Hill, New York, 1977.
8. Z. Tadmor and C. G. Gogos, *Principles of Polymer Processing*, Wiley-Interscience, New York, 1973.
9. M. Shida and L. V. Cancio, "Prediction of HDPE Processing Behavior from Rheological Measurements," *Polym. Eng. Sci.*, *11* (2), 124–128 (1971).
10. R. N. Shroff, L. V. Cancio, and M. Shida, "Prediction of Properties and Processing Behavior of LDPE," *Mod. Plastics*, 62–64 (December 1975).
11. V. M. Nadkarni and J. P. Jog, "Injection Molding Semicrystalline Polymers," *Plastics Eng.*, 37–40 (August 1984).
12. D. G. Brady, "The Crystallinity of Poly(phenylene sulfide) and Its Effect on Polymer Properties," *J. Appl. Polym. Sci.*, *20*, 2541–2551 (1976).
13. D. G. M. Wright, R. Dunk, D. Bouvart, and M. Autran, "The Effect of Crystallinity on the Properties of Injection Molded Polypropylene and Polyacetal," *Polymer*, *29*, 793–796 (1988).
14. S. S. Katti and J. M. Schultz, "The Microstructure of Injection-Moulded Semicrystalline Polymers: A Review," *Polym. Eng. Sci.*, *22*, 1001–1017 (1982).
15. E. S. Clark, "Morphology and Properties of Injection Molded Crystalline Polymers," *Appl. Polym. Symp.*, *24*, 45 (1974).
16. H.-J. Ludwig and P. Eyerer, "Influence of the Processing Conditions on Morphology and Deformation Behavior of Poly(Butylene Terephthalate)," *Polym. Eng. Sci.*, *28* (3), 143–146 (1988).
17. R. Th. Steinbach, "How the Crystallization of Nylon Affects Processing and Properties," *Mod. Plastics*, 137 (September 1964).
18. J. H. Southern and R. S. Porter, "The Properties of Polyethylene Crystallized under the Orientation and Pressure Effects of a Pressure Capillary Viscometer," *J. Appl. Polym. Sci.*, *14*, 2305–2317 (1970).
19. G. Capaccio and I. M. Ward, "Ultrahigh Modulus Linear Polyethylene Through Controlled Molecular Weight and Drawing," *Polym. Eng. Sci.*, *15*, 219–224 (1975).
20. S. Kavesh and D. C. Prevorsek, U.S. Patent No. 4422993 (1988).

Index